SpringerBriefs in Molecular Science

History of Chemistry

Series editor

Seth C. Rasmussen, Department of Chemistry and Biochemistry, North Dakota State University, Fargo, ND, USA

More information about this series at http://www.springer.com/series/10127

J. N. Campbell · Steven M. Rooney

A Time-Release History
of the Opioid Epidemic

 Springer

J. N. Campbell
Independent Scholar
Spring, TX, USA

Steven M. Rooney
Independent Scholar
Irving, TX, USA

ISSN 2191-5407　　　　　　　ISSN 2191-5415　(electronic)
SpringerBriefs in Molecular Science
ISSN 2212-991X
SpringerBriefs in History of Chemistry
ISBN 978-3-319-91787-0　　　　　ISBN 978-3-319-91788-7　(eBook)
https://doi.org/10.1007/978-3-319-91788-7

Library of Congress Control Number: 2018941231

Printed on acid-free paper

This Springer imprint is published by the registered company Springer International Publishing AG part of Springer Nature
The registered company address is: Gewerbestrasse 11, 6330 Cham, Switzerland

Though bitter, good medicine cures illness.
Though it may hurt, loyal criticism will have
beneficial effects.
—Sima Qian, Han Chinese Historian
(c. 145–86 BCE)

For Nik and her elegant deductions

Preface

History, Stephen said, is a nightmare from which I am trying to awake.
—James Joyce, *Ulysses*, 1922 [1]

Many a comic book poses the question: *what if* you could go back in time and find the exact point where some major catastrophe originated, could you make a difference by altering it? Marvel's *Days of Future Past* or DC's *Flashpoint Paradox* serve as examples of what we create for ourselves when we go back in history and change something in order to suit our own needs. Would it make a difference? These scenarios present a conundrum for those perceived to be so powerful because, as the theoretical physics goes, if you change one thing, you change everything. These stories are like the search to be modern; something that seems like a great plan, but questions abound. What other ripples does it generate when we devote ourselves to all the wonderful promises and innovations that come with it?

Then, we might ask: Where in the long history did opioids and their influence go wrong? In the media, we hear regularly that America is in the midst of a recent crisis because, since the 1990s, big pharmaceutical companies, distributors, and even drugstore chains are pumping out too many of these products. Doctors are also complicit as their over-prescription has driven people from all walks of life to addiction. The government remains powerless even though it has legislated on similar issues for over a century. In addendum, what of the academic departments in biochemistry, chemistry, physiology, and in the medical schools of the country, are they complicit? As this brief will argue, it is not one person, one entity's fault, one type opioid, per se, and there is no one point in history where all this devolved. The modern fully synthetic opioid possesses a literal time-release capability as it enters the body; but, these drugs, which have always exercised power once ingested, also have had the same effect historically. That is, their evolution, slowly, carefully, and frequently subtly, has also unfolded progressively over time. To put it another way, the current crisis when examined from a broader set of historical evidence over a wide expanse of time, assists us with understanding how the USA arrived at such a crossroads. What a macro-approach in this monograph posits is that we can deliver perspective by returning to the beginning, which will assist in adequately tracing the origins of this paradox of pain relief.

The chemical story or treatment, if you will, behind these drugs that we examine in *A Time-Release History of the Opioid Epidemic* is dedicated to the chemists and the staffs in the laboratories that were trying to solve an age-old issue—relieving pain. We have heard through news agencies, doctors, and other specialists that opioids are the root of all sorts of social maladies and that a culture war in America is raging: The employed becoming unemployed, slowly destroying neighborhoods, wrecking families; and yet, they have become central to mainstream healthcare delivery for those seeking pain relief. The media has done much to publicize and demonize the numbing (a term used in the nineteenth century) *opioid effect*. However, we have seen little in the way of scholarship from the chemical community to try and explain its origins; other than highlighting parts of its history and relating stories of addiction which do little except to frighten us with tales of misery and woe. If this truly is an epidemic, then we as a collective must seek to understand the causes of this crisis in both the distant and the recent past.

Spring, TX, USA   J. N. Campbell
Irving, TX, USA   Steven M. Rooney

Reference

1. Joyce J (1986) Ulysses, 1st edn. Vintage, New York, p 158

Acknowledgements

In our last brief, *How Aspirin Entered Our Medicine Cabinet*, we set out to produce a highly readable piece of scholarship that drew not only from chemistry and science, but a variety of fields that engaged socioeconomic history, ethnography, psychology, material thought and culture, and architecture and the built environment. What we found at the center of this modern story was a complex synthetic medicine that was born during the Industrial Revolution and accelerated in usage during the period of the Great War. Repeatedly, nations across the West turned to aspirin to solve maladies that they did not understand. They blundered through extreme body counts racked up by bullets and a Great Pandemic that produced unintended consequences. After so much destruction and conflict at the end of the Second World War, a host of chemists, doctors, pharmacists, drug companies, government entities, and advertising firms combined to wage what became known as the "aspirin wars" during the 1950s through the 1970s.

Though American society evolved and matured, digressed and faltered, aspirin's chemistry remained. In its purest form, aspirin was just that, aspirin. Nevertheless, it was far from just another member of the medicine cabinet. Thus, its chemical construct stood silently waiting while trials and cause/effect attempted to liberate its secrets. The innovations in the chemistry laboratory were integral in making aspirin a wonder drug, but it took those studies and manipulations to solve the riddle of things like heart attacks and strokes. Aspirin was, once the correct dosage tested in trials by scientists and doctors alike over successive generations, a drug that could enter mainstream use, without debate over how it could be harmful.

Our next brief, *A Time-Release History of the Opioid Epidemic* runs along a much different arc than aspirin. It seems to be the anti-aspirin, so to speak. To put it another way, if aspirin could be billed as the utopia, then opioids appear to range into the realm of something dystopian, especially when abused. This brief in long-form, spotlights what has been described by the media and the federal government as an opioid epidemic. In 2018, the USA is embroiled in a highly emotional and volatile situation that is tearing families and communities apart. As historians of science, our task is to enliven chemistry by interpreting the past. Inspired by science and technology studies (STS), we acknowledge that politics and social concerns cloud current discussions about history and create anachronisms. The opiates of the nineteenth century are not the opioids of today; however, their ecological development and influence are inextricably linked to their subsequent descendants. Over the past 200-plus years, claims of the arrival of a panacea occurred in a cyclical fashion. What we can do is analyze those patterns is explore the origins of these drugs in order to figure out what role the chemistry laboratory has played in the making of such a horrendous situation. The lab holds the key moving forward. Part science and social process, our story pertains to humanity; after all, it is individuals who built the laboratory and produced the opioids that emerged from it.

While on the subject of humans, we want to extend our gratitude to several significant people who have inspired, read, and commented on our work. They include: Michael Wintroub and Helen Mialet, whose scholarship concerning heterotopias and ethnographic approaches to all kinds of *laboratories* has been seminal to our thinking. Marcia Meldrum, Katharine Neill Harris, and Angela Kilby own work in the fields of addiction and healthcare delivery have been key to our thinking about this addiction crisis and its development. Dr. Bruce Elleman of the Naval War College reminded us about key events associated with the Opium Wars. Permission and research concerning images was assisted with access to the Lanman & Kemp Collection by Lucas Clawson of the Hagley Museum and Library in Wilmington, Delaware; Hillary S. Kativa, Curator of Photographic and Moving Image Collections at the Othmer Library of Chemical History in Philadelphia, Pennsylvania; Ilhan Citak, Special Collections, Lehigh University in Bethlehem, Pennsylvania; and Lynne Farrington, Senior Curator, Special Collections at the Kislak Center at the University of Pennsylvania. Dr. William C. Mayborn of Boston College's Woods College of Advancing Studies provided valuable expertise concerning international relations and edited the final draft of the monograph.

Special thanks to Seth Rasmussen of North Dakota State University, who is our fabulous editor for the Springer Briefs in Molecular Science. Seth knows when to give a tug on the reins or when to push the button to turn us loose. Again, he staked us by offering his support for two fellows that enjoy science, history, and writing about these topics with great passion. Our relationship with him over these past two years has been integral to our own partnership. We greatly appreciate his guidance and example. Likewise, the support from our editor at Springer, Sofia Costa, has also been essential. She has supported both of us through two projects, and for that we are grateful. Last, but not least, to both our families and friends who inspire, remind, and support our scholarship efforts; especially, Josephine who possesses a keen eye for syntax and passive voice. Our partners, Nik and Teri, are always there for both of us, and for that we are eternally grateful; chats with Jackson remind us of the blessings of youth. It is not trite to say that we could not in any way, shape or form do it without them.

Contents

1 Introduction: Chemical Heterotopias 1
  1.1 An Imperfect Resemblance 1
  1.2 Defining a Heterotopia: The Laboratory Application 3
  1.3 Scope of Current Volume 4
  References 8

2 Part One: Alkaloid Heterotopias 9
  2.1 The Roots of the Modern Opioid Empire 9
    2.1.1 Lavoisier's Vision Obscura 9
    2.1.2 Sertürner's Best Friend 11
    2.1.3 Gay-Lussac's Stock Tip for the Ages 18
    2.1.4 The First -Oid's Makeup 21
    2.1.5 The Valley of Chemicals 22
  2.2 Cooks as Chemists: America's Test Kitchen and the First
    Opioid Addiction Crisis 26
    2.2.1 Just What the Doctor Didn't Order 26
    2.2.2 Nostrum Republic 27
    2.2.3 Kiss the Cook 29
    2.2.4 Eat Me, Drink Me 32
    2.2.5 Doctors Without Borders 34
  2.3 What the Networked Industrial Laboratories Produced:
    A Heroic Decision? 37
    2.3.1 American Backwater 37
    2.3.2 Liebig's Dream 39
    2.3.3 Lab Ethnography 40
    2.3.4 Chemical Nexus: A Framework 44
    2.3.5 The Learners Become the Masters 46
    2.3.6 Bayer Goes Public 48
    2.3.7 A Chemical Rising Star 49
    2.3.8 Dreser's Dystopian Dream 51

      2.3.9   Follow the White Rabbit 53

      2.3.10 Gateway to a Synthetic Age 55

  References 56

3 Part Two: Synthetic Opiate Heterotopias 59

  3.1 The Age of Synthetics 59

      3.1.1   A New Laboratory of Progress 59

      3.1.2   The Social Drug Network......................... 61

      3.1.3   How They Made Heroin Sound Healthy 68

      3.1.4   The Costs of Slaying the Dragon: An Orwellian Tour

              of Therapeutic Reform 74

  3.2 Roots of Modern American Big Pharma Takes Hold 82

      3.2.1   The State of the Chemical Teutonic Nexus, c. 1914..... 82

      3.2.2   An Alkaloid Empire, For Sale: The 25 Minutes that

              Saved Merck 88

      3.2.3   War as Catalyst 92

  3.3 Designing Numbness: Not Your Father's Opiates 100

      3.3.1   Lab Coats of Grey Flannel: Dawn of the Opioid 100

      3.3.2   Risk-Averse Chemistry: Nazi Drugs and Massengill

              Deaths... 102

      3.3.3   The Education of the Company Man (and Woman) 105

      3.3.4   The Man Who Knew Better: The Case of Dr. Small 107

      3.3.5   The Last Great Advocate of Reason Falls 114

  References 118

4 Epilogue: Opioid Heterotopias 121

  4.1 Sackler's Ghost: The Tale of OxyContin's Failed Chemistry..... 121

  4.2 Time-Release: The Story Continues 127

  References 129

Index 131

About the Authors

J. N. Campbell is an independent scholar and is the co-author with Steven M. Rooney of the brief, *How Aspirin Entered Our Medicine Cabinet* (2017), also published by Springer. He has also written for the *International Journal of the History of Sport* and *Reviews in History*, and serves as a freelance writer for *Good Grit Magazine*. He received two M.A. degrees from the University of Kentucky and the Parson School of Design in New York City. He lives in Houston, Texas.

Steven M. Rooney is an instructor of chemistry in the Department of Science at Tarrant County Community College in Fort Worth, Texas, and also teaches courses at the Dallas County Community College North Lake Main Campus in Irving, Texas. After starting his career in research, he fervently believes in getting students interested in science by creating a daily classroom forum that fosters individual development. Rooney received an M.S. in Chemistry from the University of Missouri-St. Louis. He is the co-author with J. N. Campbell of the brief, *How Aspirin Entered Our Medicine Cabinet* (2017), also published by Springer. He lives in Dallas with his wife Teri.

Abstract

A Time-Release History of the Opioid Epidemic takes the reader on a chemical journey by following the history for over two centuries of how an opiate became an opioid, thus spawning an empire and a series of crises. These imperfect resemblances of alkaloids are both natural and synthetic substances that, particularly in America, are continually part of a growing concern about overuse. This seemed an inviting prospect for those in pain, but as the ubiquitous media coverage continues to lay bare, the levels of abuse point to the fact that perhaps an epidemic is upon us, if not a culture war. However, this is not the focus of this volume. Rather, we are seeking answers to how and why this addiction crisis transpired over two hundred years of long development. Thus, utilizing a long lens across time and space, this brief examines the role that the chemistry laboratory played in turning patients into consumers. By utilizing a host of diverse sources, this study seeks to trace the design and the production of opioids and their antecedents over the past two centuries.

From the isolation and development of the first alkaloids with morphine that relieved pain within the home and on the battlefield, to the widespread use of nostrums and the addiction crisis that ensued, to the dissemination of drugs by what became known as Big Pharma after the World Wars; and finally, to competition from homemade pharmaceuticals, the progenitor was always, in some form, a type of chemistry laboratory. At times, the laboratory pressed science to think deeply about society's maladies, such as curing disease and alleviating pain, in order to look for new opportunities in the name of progress.

Despite the best intentions, opioids have created a paradox of pain as they were manipulated by creating relief with synthetic precision and influencing a dystopian vision. Thus, influence came in many forms, from the governments, from the medical community, and from the entrepreneurial aspirations of the general populace. For better, but mostly for worse, all played a role in changing forever the trajectory of what started with the isolation of a compound in Germany. In the vein of science and technology studies (STS) in a rousing new long-form narrative, that even broadens the definition of a laboratory, we carefully probe the origins of this complicated topic.

Keywords Opioids · Alkaloids · Laudanum · Morphine · OxyContin Heroin · Heterotopia · Laboratory · STS

Chapter 1
Introduction: Chemical Heterotopias

1.1 An Imperfect Resemblance

All across the meadows, many poppies blossomed, and that were so hypnotic and brilliant in color they nearly dazzled Dorothy's eyes. "Aren't they beautiful?" the girl asked her companions, as she breathed in the spicy scent of the big, bright flowers.

—*The Wonderful Wizard of Oz* by L. Frank Baum (1900) [1]

The year 1963 saw the essence of speed and power on the first Saturday in May as the horse Chateaugay won the Kentucky Derby on the track. That summer, the release of *Jason and the Argonauts* featured the ground-breaking work of the father of stop-motion, Ray Harryhausen on the screen. Last but certainly not least, November brought tragedy as President John F. Kennedy was assassinated while in his motorcade on Elm Street in Dallas, Texas. Amidst the triumph, the creativity, and the mourning, the National Institute of Mental Health Addiction Research Center sponsored a piece that impacted the future of pain relief. Abraham Wikler, William R. Martin, Frank T. Pescor, and Charles G. Eades as a consortium published in *Psychopharmacologia* a paper concerning the "Factors Regulating Oral Consumption of an Opioid (Etonitazene) by Morphine-Addicted Rats" [2]. Despite the lengthy title it was nonetheless a fascinating study. Possibly the most interesting aspect, however, was the first footnote at the bottom of page 55. It read [2, p. 55], "In this paper, the term, "opioid," is used in the sense originally proposed by Dr. George H. Acheson (personal communication) to refer to any chemical compound with morphine-like properties" (Fig. 1.1).

With that attribution, the opiate became an opioid. Although seemingly nominal, this appellation, a process in itself, created the notion of permanence. The suffix -oid, from Greek, meaning resembling or like, is interestingly enough, like its predecessor the alkaloid, just a façade.[1] The implication when employing such an ending is that the adjective or noun is an imperfect resemblance to what is indicated by the preceding element (appropriate term considering this is a chemical subject) [3]. Therefore, the

---

[1] An -oid originates from Latinized form of Greek *oeides*, from *eidos* "form," related to *idein* "to see", *eidenai* "to know" literally "to see," from PIE *weid-es-*, from root *weid-* "to see, to know".

© The Author(s) 2018
J. N. Campbell and S. M. Rooney, *A Time-Release History of the Opioid Epidemic*,
SpringerBriefs in History of Chemistry, https://doi.org/10.1007/978-3-319-91788-7_1

Fig. 1.1 Opiologia ad mentem, Academiae Naturae Curiosorum by G. W. Wedel, c. 1674. Photo courtesy of the Roy G. Neville Collection, Othmer Library, Chemical Heritage Foundation

word opioid, like an anthropoid, a planetoid, an android, a hyoid or to use a slang term, a zomboid, is something incomplete.[2] Even the word tabloid, which was coined by the Burroughs Wellcome & Company of England in 1884 when they trademarked it to promote their new line of tablets, translates to something that is compact or smaller than usual (Fig. 1.2) [4].[3] To put it another way, an -oid, and thus an opioid, is similar, yet unto itself. It is an opiate, but also not an opiate; it is something like, but also wholly different.

Similar to this suffix, human beings can also be understood as imperfect or incomplete resemblances of their true selves. When positioned on a map, humans can be

---

[2]An interesting note, the primary function of the hyoid bone, as an irregular bone, is to serve as an anchoring structure for the tongue. The bone is situated at the root of the tongue in the front of the neck and between the lower jaw and the largest cartilage of the larynx, or voice box. In forensic chemistry, if during an autopsy a corpse is discovered to possess a broken hyoid it can be surmised that strangulation was a possible cause of death.

[3]*Tabloids* have a deep connection to the laboratory and pharmaceuticals. Sir Henry Wellcome was such an excellent salesman in Britain that he wanted to create a brand around the name tabloid because it was connected to alkaloids which were, as this study will examine in Sect. 1.2, so integral to the development of opiates and opioids. He expanded the name beyond just tablets by producing everything from teas to medicine chests that famous explorers could take on their expeditions. Of course, tabloids as a term eventually became so enmeshed in the lexicon that eventually a judge ruled that Burroughs Wellcome could no longer trademark the term. Today, it is associated primarily with newspapers and magazines that sensationalize everything from the Kardashians to Bigfoot sightings.

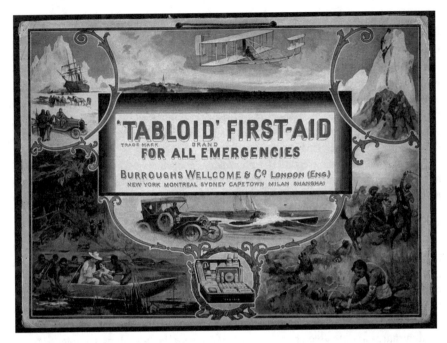

Fig. 1.2 Tabloid medicine chest made by Burroughs Wellcome, c. 1909. Photo courtesy of the Wellcome collection

seen as both dispersed and centered, depending on the networks of people in which they identify and come into contact. These pathways could be literal, as in asphalt or concrete or figurative, as in connected by fiber or in a virtual world. Regardless, we are part of inner and outer connected networks of people. Continually, those synapses of communication are bombarded with how to cope with decisions, sometimes seemingly unimportant ones that were made in the near and distant pasts. Whether we acknowledge it or not or revel in our own magnificence or not, we are both the dreamers and the damned, mirrors of our best and worst sides simultaneously. That is not only poetry, nor just science, per se, but it is the way this world operates.

1.2 Defining a Heterotopia: The Laboratory Application

To this end, we live in what the social historian and theorist Michael Foucault interprets as a heterotopia, like an -oid; a place that is and is not exactly what it seems, but something that ebbs and flows depending on one's place [5].[4] Foucault pushed

---

[4] See Wintroub M (2017) *The voyage of thought: navigating knowledge across the Sixteenth-Century world*. Cambridge University Press, Cambridge, 17, 23–24; A heterotopia also exists in medicine, for instance, when a particular tissue type is located at a non-physiological site, but also coexists with

us to see that we do not live in a void; rather, our structures accumulate objects, ideas, and spaces with time. Like museums and archives, time never stops building towards its own immovable summit. In contrast, almost simultaneously, we are also like the fairgrounds or the carnival. As humans we move, pick up, and transmit, along with transferring our ideas and objects into our "trunks," which serve as windows into where we have been. We can be itinerants that pick up our tent poles and nomadically shift from place to place, always mindful of where we are headed, and forgetting where we have been. Thus, as Foucault delineated, our spaces can be temporary or they can be continual places of accumulation.

Within this network we find the chemistry laboratory, which, as this piece will argue, is a place of constant movement (including of chemicals, people, and between labs). Yet, the lab is also founded on the type of science (unique among its peers in its arc of development) that has compounded since the nineteenth century [6, p. 5].[5] The *lab* (which is a shortened version of the word, *laboratory*) has served as an intersection for different research fields over the centuries, while having the ability to translate problems from one to the other. As a chemical heterotopia it is a vehicle that serves as a veritable cross-reference guide. The lab, how it functions and what it produces, can be both fixed and mobile. Furthermore, it has never been just a place full of white-clad chemists surrounded by technical equipment. What it continues to be is a place of wonder, excitement, and where chemical dreams come true; or a venue of nightmares, unintended consequences, and where the best chemical intentions go awry—a heterotopia.

1.3 Scope of Current Volume

This is the chemical story of a health crisis like and unlike any other; a true -oid, if there ever was one. Our subject has certain parameters, which we will redefine. The opioid addiction crisis especially when viewed through the prism informed by the media has become a tale of reductionism, like the reader who skips to the end of a mystery novel. The path to pain relief is more complicated, so with synthetic precision, like our subject was intended, we will delineate how the laboratory is anthropomorphic—the laboratory is filled with chemists that create, test, and synthesize. Rather than drawing simplistic conclusions and utilizing a host of assumptions, we would prefer to examine the deep historical roots of this chemical breakdown. From the separation of early alkaloids in the nineteenth century through the influence of the

---

the original tissue in its correct anatomical location. Even more fascinating is that heterotopias exist within the brain and are often divided into three groups: subependymal heterotopia, focal cortical heterotopia, and band heterotopia. Thus, we can use our heterotopias to think about heterotopias in cultures and societies!

[5]Benedict Anderson in his oft-cited work, *Imagined Communities* [6], uses the phrase *sui generis*, which translates to "unique in every way possible," to describe the construction techniques used in building nations. Since, the laboratory was an extension of the state that at times provided funding it has an appropriate assignation here.

German labs of the pre-1939 period, and finally to the prodigious leaps in potency all the way to the present, we seek to use the chemistry lab as a means to examine this supposed panacea, in its many forms throughout history.

The lab, over the course of its own development and maturation, is a heterotopia where, depending on its period and place can be like the museum (a monolith of collection) or the fairgrounds (a movable target). Drawing inspiration from the work of Bruno Latour and Steve Woolgar's work, *Laboratory Life: The Construction of Scientific Facts* (1978), and the science and technology studies (STS) that emanated from subsequent ethnographic and anthropological research, our objective is to apply similar principles of investigation to the chemical origins of opioids [7–9].[6] This brief seeks to answer several important questions. Namely, how did the chemists and their staffs reconcile the work they engaged in when paired against ideas about treating pain and the corrupting influences of business, advertising, and governmental policy-making? Even more so, how did they conceive of what they were doing across a broad network of disciplines including biochemistry and clinical medicine that frequently contradicted one another? For that matter, how did the evolving design and definition of our concept of a lab, influence the process of creating first opiates, and then, under the auspices of corporate capitalism, opioids? Thus, understanding how the laboratory could produce a chemical heterotopia that would assist in turning patients into consumers can inform us about the current crisis. Opioids, as a slow-moving locomotive that gained a head of steam after nearly two centuries of development, will be the focus of this volume.

In 2018, with over 200,000 opioid-related deaths since the beginning of the century, the United States, and every other nation on the planet is susceptible to something like the current opioid crisis [10, 11].[7] How is this possible? How can a supposed hegemonic nation in possession of such wealth and technological savvy, be on the cusp of an utter disaster such as this one? Most importantly, how did we arrive at this point in *our* history where healthcare delivery, backed by the powerful pharmaceutical companies, which employ some of the best chief chemists, be for lack of a better term, so myopic? Answers always spotlight the manufacturers, wholesalers, distrib-

---

[6]Latour and Woolgar were not the first progenitors of the STS scholarship. That started in the 1960s with the publishing of Thomas Kuhn's, *The Structure of Scientific Revolutions* (1962), which examined the underlying intellectual shifts behind the rise of European innovations in the sixteenth and seventeenth centuries. Taking cues from historians of science that were now joined in departments at universities by philosophy professors, STS was built on examining technology in its social and historical context. These historians guffawed at technological determinism, a doctrine that promoted mindlessness by the public through acceptance of an Orwellian world. At the same time, others began to develop similar contextual approaches to medical history. Since then, the movement has added juried journals, conferences, and a host of other outlets including the development of subfields like Tecnoscience and Tecnosocial.

[7]All sorts of articles and news agencies are reporting numbers and forecasting death rates. Sources need to be scrutinized in order to draw the proper conclusions. The fact of the matter is no one really knows moving forward what will occur. One of the more compelling charts we found was *The Opioid Epidemic* (2017), produced by Andy Brunning in collaboration with *Chemical and Engineering News*. Brunning's website, http://www.compoundchem.com/ is also an interesting visual source concerning chemical compounds.

utors, and even the drug store chains, and themes gravitate towards pharmaceutical greed, the influence of drug lobby groups or the clipped wings of the Drug Enforcement Agency. Several thousands of pages have already been written about opioids, and most begin by sensationally looking at only the recent past, *The Doctors Who Started the Opioid Epidemic*, or *The One-paragraph Letter From 1980 That Fueled the Opioid Crisis* (for a discussion of this source see Sect. 4.4.2) [12–14].[8] Yet, not one of them attempts to examine the long history of the laboratory and the course that this crisis has taken; and for that matter, not one of them utilizes a meta-ethnographic approach to the laboratory.

This volume begins in Part One by examining the eighteenth and nineteenth century laboratory heterotopias prior to and on the eve of industrialization. Staffed by a cadre of pharmacists and their apprentices and opposed by gentlemen scientists and professors at universities, it was through a combination of their efforts that the first opium experiments gave way to the first alkaloids—the building blocks of morphine, codeine, and opioids. Revolutionizing pain, the opium poppy particularly in China, incited wars and cracked open that aged Empire to outside influences through a drug war. Opium use accelerated in America, although in the form of laudanum, which contained a tincture of the juice and were known in the trade as patent medicines. The first opioid addiction crisis ensued as everyone from the government to the field of chemistry debated its usage. These compounds were developed in "labs," sometimes in household kitchens, that did little to regulate safety. Chemistry laboratories attempted to replicate alkaloids into morphine-like substances which were finally developed after German unification in 1871. These new industrial labs were painfully aware of what their products could do to society, and they attempted to govern themselves accordingly. As streamlined heterotopias, these spaces always had one eye on the past and the other on the future. The new medications produced from tar-based synthetics created the modern pharmaceutical industry and quickly outpaced the rule of law on the eve of the Great War.

In Part Two, we see that technology and the state continually emerged after 1865 from modern and then global conflicts with a renewed sense of organization and development; and so did the chemistry lab, which had quite a run before and during the Second World War. Examining the effects of the professionalization of healthcare delivery combined with the creation of bigger pharmaceutical laboratories in the postwar era will be our focus as lab architecture influenced production through a network of collaboration. In turn, this led to a host of new pain-killing products, namely, the opioid. Now with new trials and algorithms, the lab accelerated its output into what became a pharmaceutical heterotopia. In the Epilogue, as innovation led to disillusionment, the United States legislative campaigns failed to rein in production, and the laboratory was redefined once again as street drugs mixed with pharmaceutical opioids like never before. Thus, a fully synthetic compound, which had evolved slowly over its history since the late eighteenth century, would be unleashed by chem-

---

[8]An example of a recent article in the *Washington Post's* section entitled, *Retropolis*, that attempts to link the past with the present is a piece by journalist Nick Miroff. For the record, he makes no mention of the role of the chemistry laboratory in the development of opioids.

istry and time. With control firmly in the hands of Big Pharma, they manipulated and altered drugs such as OxyContin leading to abuse and even death. Poised to introduce an antidote to the current crisis we argue it is incumbent on the heterotopia to not just innovate, but to pursue a different path like never before [15].[9]

As our study moves through different laboratory heterotopias, we will try to strike an important balance that reflects the unfolding of time-release across history. To herd opioids into one particular conglomerate and tag them as such is also not a productive endeavor. As this work will show, Morphine is not Pethidine, and Methadone is not OxyContin. Each opioid has its own history, its own arc of development; and like the chemists and laboratories behind them, they have much to say about the cultures that produced them. In other words, they simultaneously stand on their own; and at the same time comprise a larger history of systematic development across time and space. Like the proverbial inebriate who only looks for her or his keys where the light shines under the lamppost, so it is with the scholarship and media-driven circus that is the opioid epidemic.

A Time-Release History of the Opioid Epidemic traces the story of the rise of this controversial class of pharmaceuticals. In an era of fixed combination medications, they became the dystopian drug of the nineteenth century, a point of hope and then declared illegal in the Progressive Era, and in the postwar era the personification of a new modern corporate design. There was a time when opioids were the last best hope for pain relief and in 1965 the federal government made the legislative distinction between prescription drugs and outlawed narcotics. Yet, something happened in American culture, and it was not just affluence, the invention of the mid-life crisis, or the inability by the next generation to cope with pain. The key to opioids and their destructive qualities reside in the networked laboratory that was industrialized during the late nineteenth century. Certainly, the pushing of opioids by doctors and prescription drug companies in the recent past is important to understand; but the argument in this monograph reaches further back. There is a larger story for why this grouping of drugs invites the retiree who spent a lifetime climbing up and down ladders on an offshore oil platform or the football player who blew an ACL in his leg during his last high school game or the stay-at-home mom who just had elbow surgery; they all are susceptible to addiction. However, this brief is not about them; but they might do well to read it.[10] There are plenty of monographs, articles, and essays that examine the economic depression over what has transpired in Ohio,

---

[9]We want to acknowledge that interpreting events of the recent past is in our estimation to be fraught with peril. In a sense, releasing interpretations over-time, like our subjects, does have its benefits. Thus, getting away lends perspective. Though time does not necessarily heal all wounds; it does offer panoramic views of the currents of the past. One of the many benefits of the study of World History, espoused by the historian Jerry Bentley, presents these types of opportunities in which larger expanses of events can change the ways in which people view current events from history.

[10]If you are interested in reading about the recent sociological and personal impact of opioids you might consult the popular work of Sam Quinones in *Dreamland: The True Tale of America's Opiate Epidemic* (2015) or J. D. Vance's memoir, *Hillbilly Elegy: A Memoir of a Family and Culture in Crisis* (2016). The non-profit efforts of S.A.F.E. (Stop the Addiction Fatality Epidemic) are also attempting to prosecute this crisis, www.safeproject.us.

Kentucky, and West Virginia. Rather, this is a chemical story of how the laboratory as a heterotopia became both the greatest progenitor of scientific therapeutic progress, while at the same time produced a dystopian drug of imperfect resemblance.

References

1. Baum LF (1900) The wonderful wizard of Oz. George M. Hill Company, Chicago
2. Wikler A, Martin WR, Pescor F, Eades CG (1963) Factors regulating oral consumption of an opioid (etonitazene) by morphine-addicted rats. Psychopharmacologia 5:55–76
3. -oid definition (2017) http://www.dictionary.com/browse/-oid. Accessed 17 Oct 2017
4. Larson F (2009) An infinity of things: how Sir Henry Wellcome collected the world. Oxford, Oxford
5. Foucault M, Miskowiec J (trans) (1986) Of other spaces. Diacritics 16(1):22–27
6. Anderson B (2006) Imagined communities: reflections on the origin and spread of nationalism. Vergo, New York
7. Latour B, Woolgar S (1986) Laboratory life: the construction of scientific facts. Princeton University Press, Princeton
8. Gross M (2008) Give me a laboratory and I will raise a discipline: the past, the present, and the future of politics of laboratory studies in STS. In: Hackett EJ, Amsterdamsk O, Lynch M, Wajcman J (eds) The handbook of science and technology studies, 3rd edn. MIT Press, Cambridge, MA, pp 279–318
9. Cunningham A, Williams P (eds) (1992) The laboratory revolution in medicine. Cambridge University Press, Cambridge
10. Barnett M, Olenski AR, Jena AB (2017) Opioid prescribing by emergency physicians and risk of long-term use. N Engl J Med 376:663–673
11. Brunning A (2017) The Opioid Epidemic. Chemical and Engineering News. 12 June 2017 https://cen.acs.org/content/dam/cen/95/24/09524-scitech3.pdf. Accessed 1 Oct 2017
12. Offit P (2017) The doctors who started the opioid epidemic. The Daily Beast. 1 April 2017 https://www.thedailybeast.com/the-doctors-who-started-the-opioid-epidemic. Accessed 2 Aug 2017
13. Zhang S (2017) The one-paragraph letter from 1980 that fueled the opioid crisis The Atlantic. 2 June 2017 https://www.theatlantic.com/health/archive/2017/06/nejm-letter-opioids/528840/. Accessed 1 Aug 2017
14. Miroff N (2017) From Teddy Roosevelt to Trump: how drug companies triggered an opioid crisis a century ago. 17 October 2017. https://www.washingtonpost.com/news/retropolis/wp/2017/09/29/the-greatest-drug-fiends-in-the-world-an-american-opioid-crisis-in-1908/?utm_term=.bf663795f478. Accessed 1 Nov 2017
15. Bentley J (2007) Why study world history? World History Connected. 5.1: 19 pars. http://worldhistoryconnected.press.illinois.edu/5.1/bentley.html. Accessed 14 Oct 2017

Chapter 2
Part One: Alkaloid Heterotopias

2.1 The Roots of the Modern Opioid Empire

It took them only an instant to cut off that head and

a hundred years may not produce another like it.

—Joseph Louis Compte Lagrange, French mathematician and astronomer, commenting on the murder of Lavoisier, c. 1794 [1]

2.1.1 Lavoisier's Vision Obscura

What went through the mind of Antoine Lavoisier right before the guillotine descended upon his neck, cutting off so much promise as one of the founders of modern chemistry? We cannot help but wonder. Perhaps he was thinking about the experiments he would never get to carry out; maybe he was pondering why his efforts to fight the adulteration of tobacco or to found a Royal Commission on Agriculture had fallen short; or odds on, he was picturing his beautiful and intellectually stimulating wife and collaborator in the laboratory, Marie-Anne, who had steadfastly defended him to the last against the Jacobian steamroller. Lavoisier had the misfortune of being on the wrong side of a revolution that became highly radicalized after the Reign of Terror. Given the opportunity, he could have proffered so much more for the field that fueled both his passion and his marriage. The latter was an interesting pairing because his wife was so integral to everything he did in his work and as a *colleague* in a state of the art laboratory that he constructed for the next generation of chemists [2].

The chemistry laboratory, a heterotopia that would see vast changes over the course of the nineteenth century, evolved steadily from the early modern into the modern world by adding new architectural features, the latest glassware and heating devices, and diverse personnel that specialized in different forms of chemical practices, just to name a few. As previously suggested, it was a heterotopia that could ebb and flow depending on the period and whose influence evolved. Europe set the tone for chemical development, first through the work of gentlemen scientists who were

© The Author(s) 2018
J. N. Campbell and S. M. Rooney, *A Time-Release History of the Opioid Epidemic*,
SpringerBriefs in History of Chemistry, https://doi.org/10.1007/978-3-319-91788-7_2

well-connected to wealthy patrons or were part of the landed gentry. The Enlightenment assisted in levelling the playing field so-to-speak as pharmacists, university professors, and students of all-kinds vied for new vocations. Within these growing laboratories, knowledge and experimentation mattered most and pressed chemistry to move from an obscure field to the forefront of nation-building. In a way, before the revolution in the lab could take hold, a radical political one would ravage France and change the future of chemistry forever [3].

Few associated with the machinations of the First or Second Estates in France survived the period that encompassed the fall of the Ancien Regime, the Revolution itself, and the Napoleonic Era like the artist, Jacque Louis David. His skill as a court painter at the Louvre, political provocateur and associate of Robespierre, and mentor to a generation of French painters spanned over four decades before his death in 1825 [4]. Some of the most influential figures in French history sat for him, but his depiction of Antoine Laurent Lavoisier and Marie-Anne Pierrette Paulze (a former student of David) exhibits both a revolutionary marriage and an early example of how a laboratory could function under the auspices of a husband-wife team (Fig. 2.1).[1] Commissioned by the couple, the painting evokes some fascinating details such as the equipment on the table next to the seated Lavoisier, which includes a barometer, gasometer, water still, and a glass bell jar. A large round-bottom flask and a tap are on the floor to the right by the table. Yet, the most vital extension of Lavoisier's talent was the woman next to him, Madame Marie-Anne. Standing above her husband, she penetrates the viewer with her eyes, while she gently places her left arm on his shoulder, as he looks up. She was not an ordinary wife and she was no ordinary chemist [4].

The Lavoisiers were progenitors of the revolution in chemistry that was sweeping Western Europe. At the time, few could rival their self-sponsored laboratory because it was constructed with the latest equipment and specifically for aspiring scientists who could study without the pressure associated with securing funding for their own research. Marie-Anne herself was an accomplished student not only of drawing (Fig. 2.2), but as an effective translator. She also understood chemistry and edited her husband's treatises by catching mistakes that would have otherwise sabotaged some key proofs [5]. As a scholar she inhabited a heterotopia within the lab that was rarely seen by a woman, especially in the eighteenth century. Like American First Ladies, Abigail Adams and Dolley Madison, who were partners in politics with their husbands, Marie-Anne devised and hosted salons with visiting scholars, and students as well as members of her husband's chemical circle. Not only did this advance her partner's work, but it also informed her own base of knowledge as a chemist. Lavoisier's impact on the field was cut short literally by his own politics since the Revolution incorrectly deemed him a threat. Based on his writings, which were returned to Anne-Marie with a tardy apology for murdering her husband, had he lived it is posited that their work in plant chemistry would have yielded a revolutionary discovery: alkaloids.

---

[1]David himself was connected to Robespierre and the Jacobins, but his renown as an artist allowed him to escape both the Terror and the subsequent Thermidorian Reaction (July 1794) which spelled the end of Jacobian rule. Lavoisier was killed just a month before (8 May 1794).

Fig. 2.1 Antoine Laurent Lavoisier and Marie Anne Lavoisier, Artist: Jacques Louis David c.1788. Photo courtesy of Edgar Fahs Smith Collection, Kislak Center, University of Pennsylvania

2.1.2 Sertürner's Best Friend

Instead of the Lavoisier's lab being credited with unlocking the secrets of alkaloids, the honor of discovering the chemical antecedents to opioids was not a member of the Royal Society or one of the deans of modern chemistry. The honor belongs to the innocuous Friedrich Wilhelm Sertürner. Historians have considered him to be everything from an apothecary's assistant, to a pharmacist, and even a full-fledged chemist.[2] He crossed between each of these titles over the course of a long career.

---

[2] We must tread carefully here as the use of apothecary and pharmacist are used interchangeably in the secondary literature. The two terms really do not refer to the same knowledge base. An apothecary was someone who, like a pharmacist would dispense medications after a simple apprenticeship, but would also perform diagnoses. Depending on the historical timeframe, the terms began to merge with different regions adopting them at various points. The situation became even more complex

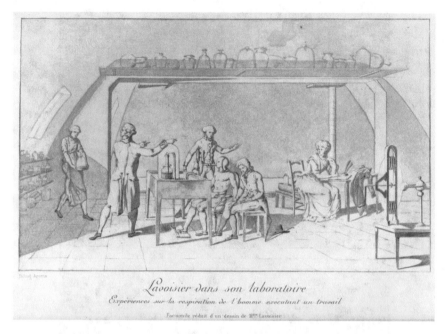

Lavoisier dans son laboratoire
Expériences sur la respiration de l'homme exécutant un travail

Fac-simile réduit d'un dessin de Mme Lavoisier

Fig. 2.2 Lavoisier in his laboratory: Experiments on breathing a man carrying out a task. Reduced facsimile of a drawing by Mrs. Lavoisier, c. 1780, from Edward Grimaux, imprinted 1888. Photo courtesy of Edgar Fahs Smith Collection, Kislak Center, University of Pennsylvania

Regardless, what is most significant is that he played a decisive role in inaugurating the age of the alkaloid. How did a virtually unknown student solve the earliest key to the modern opioid and become part of the history of materia medica?[3] And, what were the immediate effects of his discovery on the field of chemistry and drug manufacturing? The answer lies in the chemical climate that would set the tone for the field as the rest of the nineteenth century unfolded. Sertürner's moment in the history of chemistry was part of the revolution that was changing the field during the 1770s and 1780s. In a series of discoveries prior to his work in 1804 coupled with changes that were afoot in both pharmacology training and the political climate between France and the Germanic States, this nexus of events led to an explosion in the development of alkaloids [6].

The world of organic chemistry was built on a foundation of herbal medicine and also by what came to be known as animal and plant chemistry. Both were not mutually exclusive subjects. Across Europe, pharmacists were attempting to keep pace with chemists, like the Lavoisiers. During the eighteenth century these medicinal practi-

---

with the introduction of the term "chemist," which was, at least beginning in the early nineteenth century, also used to describe those that practiced university-trained pharmacology.

[3]Materia medica is Latin term for the body of collected knowledge about the therapeutic properties of any substance used in healing. Used from the 1st Century from Greece and Rome, but by the twentieth century it has now been replaced by the term pharmacology.

tioners sought to shake off the persona of medieval healers, and professionalize by incorporating the latest innovations, while maintaining a system of apprenticeship [7]. Due to university building influenced by Enlightenment thinking, pharmacology was evolving from solely a teaching exercise that passed knowledge from one generation to another to an academic profession during the second half of the eighteenth century. Who would have thought that while France and America dealt with their own respective political revolutions, all kinds of organic acids were being isolated during the 1770s into the 1780s?[4] Using older apparatus, new chemical compounds and elements were unlocked, such as carbon dioxide and nitrous oxide. Scientific journals reported these findings in print culture that could be found primarily in urban environments. Thus, regional practices could now be read and debated in the salons and coffeehouses (which served as an extension of the lab heterotopia) by those interested in emerging fields. Despite revolutionary mayhem, France emerged as the progenitor of plant chemistry. Yet, this distinction would not last. With so much upheaval, as we saw with the death of Lavoisier, the French would not be able to corner the market on chemistry. In a lively exchange of experimentation, the pharmacists across Europe combined centuries of technical skill with a new level of knowledge that was acquired through the printing of the latest in both botany and chemical training. Central to work performed within these two fields was the belief that herbal ingredients would only be present in acids [6]. That belief was about to heavily influence the discovery of morphine.

Friedrich Sertürner never got to follow in his father's footsteps in the prince-bishop engineering services located in the city of Paderborn in the western part of Germany because the old man died.[5] Born in 1783, by the age of fifteen the young Sertürner's mother could not afford to buy him a position of such station. It was an inauspicious beginning to say the least. Yet, the 1790s in the German state of Westphalia was brimming with opportunity for those on the make. Napoleon's march across Europe that would open trade through his raucous Continental System was still far off, but still the Germanic states were abuzz with scientific activity. But, not Paderborn, which despite its status as the capital of the province was not a Berlin, a Munich or an Erfurt. It was in this supposed rural backwater that we find Sertürner in 1799 beginning a four-year apprenticeship to a court apothecary [8].

For centuries those that practiced herbal medicine engaged in a veritable race to unlock the secrets of opium. A narcotic that comes from the latex resin of the immature seed pods of opium poppies it can grow naturally in diverse places such as modern China, Afghanistan, Turkey, and Mexico. Dispensed by apothecaries that were Arab and Christian alike, the poppy was manipulated by those who practiced all sorts of forms of pharmacology (Fig. 2.3). Opium contains up to 12% morphine

---

[4]Citric acid (1774), sour glucic acid (1778), and malic acid (1785) are examples of this explosion in new chemical discoveries.

[5]Friedrich Wilhelm Sertürner (1793–1841) was born into a family with deep local connections in Paderborn, which is north of Frankfurt, Germany. After his success with morphine he owned pharmacies, founded a journal, authored monographs, and worked on many other chemical projects, including the search for a cure for cholera. After receiving many accolades, he died of gout in February 1841.

and up to 2% codeine, which are the main opiate alkaloid ingredients that give it its narcotic properties. The use of opium dates back centuries to 4200 BC where capsules of poppy seeds were unearthed near ancient burial sites. Documentation proves that a poppy seed elixir was used as far back as the Byzantine Empire; though tales of its rediscovery are inaccurately attributed in 1522 to a traveling doctor named Paracelsus [9].[6] He named his concoction laudanum and used it as a painkiller. Laudanum (discussed further in Sect. 2.2) became a medical standard beginning in the eighteenth century, but the issue was that it did not provide a consistent dosage. Hence, those that could unlock the secrets of opium would receive more than a blue ribbon; instead a place in chemical history awaited them. Since the seventeenth century salts were found and reactions produced alkaline developments, but always impurities stymied a great leap forward. Interestingly enough, the laboratory heterotopia that Sertürner worked in had virtually no knowledge of this past history. And, in this case, that was extremely important.

When Sertürner began his apprenticeship, he had a diverse group of apothecaries around him, including Franz Anton Cramer, who served and advised the local court. They had connections to an herbal past, were used to serving authoritarian masters that demanded results, and possessed an institutional knowledge that included a set of time-honored traditions. The laboratory he worked in allowed for interpretation, but it still had standards that could be found in larger urban academic institutions. After his four years elapsed, testing was rigorous and though a degree was not specifically conferred, Sertürner received numerous opportunities to ply his new trade [8]. By 1804, with plant chemistry and pharmacology now part of his milieu, in his final year before obtaining a position as an assistant in a municipal pharmacy just east of Paderborn in the town of Einbeck, he had a breakthrough. What we know is that he began his scientific investigations not blindly nor due to carefree sensibilities as some scholars have suggested, but because he was part of an intellectual and chemical heterotopia that was going through a revolution in practice. Scientific methods were being overhauled, and in the wake of French chemistry, Germanic states were adopting and incorporating the past and the present into pharmacology [8]. Sertürner was part of this new generation of apothecaries which would, after the Napoleonic Period, move directly into university settings and found departments devoted to chemical studies.

---

[6]During the height of the Reformation, as mentioned, opium was supposedly reintroduced into European medical literature by Paracelsus (1493–1541). The historical issues concerning him abound. Many sources repeat fallacies about his contributions. What we do know is that he was the first to use tincture of opium (an alcohol extract of opium), but it is doubtful that he "rediscovered" opium. In terms of Paracelsus himself, his actual name was Theophrastus Bombastus Von Hohenheim and he is more properly either German, or Swiss-German, not Swiss (which is another inaccuracy), as his birthplace of Zurich, which was part of the old Holy Roman Empire at that time. He was definitely an alchemist and practiced medicine, although it is still debated if he actually had any medical degrees. He studied with astrologers, but he himself was not a practicing astrologer. In 1522, he was most certainly not a "traveling doctor", but an army surgeon and did not start practicing medicine until a few years later. At that time surgeons were not doctors, but medical craftsmen, that cut where the doctors instructed. The authors want to thank Seth Rasmussen for his insight into this subject.

Fig. 2.3 Opiologia ad mentem, Academiae Naturae Curiosorum by G. W. Wedel (Wedel G. W.) (1674) *Opiologia*, ad mentem Academiae Naturae Curiosorum, Jenae, sumptibus Johannis Fritschii. 4to (198 × 150 mm). pp. (8), 170, (2), title printed in red and black with a large engraving. Old half vellum. Georg Wedel (1645–1721) was professor of Medicine at Jena and was received into the Academia Naturae Curiosum, with the name Hercules I. He was first physician of the Duke of Weimar and the Duke of Saxony. Described as an excellent scholar and humanist, he wrote numerous books, brought out new editions of the works of older writers, and was the author of a host of disputations, consilia, responsa, paradoxa, orationes, programmata, and epistolae, written in Latin. The work includes an engraving on the title, showing an Arab apothecary in the process of preparing opium in a typical idyllic German landscape, with a river, trees and farmhouses (See, Fig. 1.1 for a closer image). Inside, the work describes the pharmacological aspects of opium, and reflects the close and longstanding relationship of Middle Eastern suppliers of opium to West. Photo courtesy of the Roy G. Neville Collection, Othmer Library, Chemical Heritage Foundation

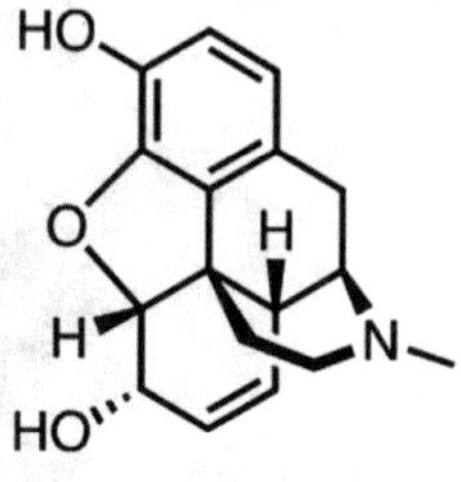

Fig. 2.4 Modern line
structure of the morphine
molecule

What exactly did Sertürner do chemically to yield such a breakthrough as morphine? What any good scientist would—he found a subject for his experiments. Well,
more specifically, he found from the family Canidae Fischer—a dog. Some evidence
points to several, but the record is unclear. We are not quite sure if this was a stray
off the street or someone's pet, but like Pavlov's famous canine, this animal would
become the first victim of a morphine overdose. Why Sertürner chose to work and
experiment with opium is unclear, but as mentioned this was a pursuit that many
chemists during the period undertook [10]. With Europe engulfed in pain due to the
outbreak of disease and warfare, many states were interested in funding research.
Similar to his colleagues in medical dispensaries across Europe, Sertürner focused
on the conundrum of whether herbal ingredients would be present only in acids.
The acidic revolution which had swept through labs back in the 1780s served as
an inspiration, and once again the herbal past intersected with a series of chemical
moments.

Through 57 iterations, he began by boiling opium down into what was called isolating morphine (Fig. 2.4) Starting with approximately eight ounces of dried opium
the substance was digested with heat and extracted several times with distilled water
until it was no longer colored. After evaporation, a translucent extract was obtained,
which became cloudy and opaque by dilution with water. The liquor regained its
transparency by either heating the solution or adding more water. This aqueous solution of the extract was saturated by an excess of aqueous ammonium hydroxide and a
greyish white substance precipitated, which formed itself in granular, semitransparent
crystals. In turn it was these crystals that were believed to be morphine. Next, while
combined with some insoluble matter and meconic acid the substances were washed
several times with water. After drying, the crystals were dissolved with a diluted
sulfuric acid solution and were recrystallized by the addition of aqueous ammonia.
Several attempts were implemented to separate the morphine from the impure material. This process did not isolate the morphine completely. The powder that resulted
was then recrystallized numerous times with a little alcohol. The morphine that was
recovered by recrystallization was now colorless indicating a reasonably pure sample
[11].

Halfway through the set of experiments Sertürner decided it was time to give
a dosage to a living being as a test. Like America's Ham the Chimp who had the
honor of being the first hominid in space in 1961, Sertürner's dog was the first to
experience the euphoria of morphine. Hazy and probably liking the sugar juice that

was incorporated, we can only imagine what it was like to watch this trial unfold [8].[7] What we do know is that endorphins are natural pain relievers that work to remove or diminish nerve signals. As a mammal's body becomes stressed, the endorphin concentration increases. Morphine has the same effect as endorphins on the nerve signals. A portion of the molecular structure of morphine mimics the molecular structure of naturally produced endorphins and produces the same pain-relieving response in the brain. In fact, a common feature to pharmaceutical agents used to create a state of euphoria or block pain signals is the β-phenylethylamine molecular component. For these compounds to be effective they must be able to cross the blood-brain barrier. Functionally, the brain is surrounded by tissue that blocks the passage of large molecules and charged ions. Nutrients can only enter by special transport mechanisms. Only small, fat-soluble molecules that have no charge will be able to pass freely into the brain tissue. Molecules like the β-phenylethylamine derivatives can pass the blood-brain barrier freely. The quicker the drug passes through the barrier, the faster the physiological response. Unfortunately for Sertürner's canine the next dosage proved fatal; thus, the age of morphine had arrived [8].

Despite his ill-regard for animals (it should be noted that animal rights groups were on the rise in Britain), by 1805 Sertürner's prospects improved; his father would have been proud. He had received the position in Einbeck, and readied his conclusions from the opium tests for publication in the Spring Edition of the *Journal of Pharmacy* which was edited by the seasoned pharmacist, Johann Bartholomew Trommsdorff. Sertürner was dipping his toes into some deep water, but he was making some prescient statements that would be debated time and time again by chemists, pharmaceutical companies, distributors, and of course, healthcare professionals. He stated in the conclusion to the 1806 publication of his work that, "the doctor no longer has to struggle with the uncertainty and vagaries of what is often complained about, he will always be able to use the remedy with the same success." This was quite prophetic considering that within two years Sertürner had helped to redefine experimental pharmacology [9]. Despite his findings, his colleagues within the fields of chemistry and medicine were less than enthusiastic. Instead of embracing not only the results of his experiments, but also honoring the laboratory training he had received in Paderborn, gatekeepers like Trommsdorff offered mixed reviews at best. Many claimed by overstating the obvious that his conclusions were acceptable, but more should be done. Undeterred, Sertürner marched on. Taking full advantage of Napoleon's Continental System, which attempted to sever British trade in and out of Europe while allowing inter-European commerce to flourish, he passed the next phase of his pharmacologist's exam before the medical college in Kassel and proceeded to found a second pharmacy in Einbeck.[8]

---

[7] It should be noted that after a second dose, the dog stumbled and became drowsy enough to induce sleep. Sertürner was able to revive the animal with a weak acetic acid. This substance was an alkaline.

[8] Napoleon was Master of Europe, but his Continental System was an utter failure. While it did limit British manufacturing, it only in the end hurt French industry. Other seaborne empires suffered under it, like the Dutch, but central Germanic States like Sertürner's did until Napoleon's defeat at Leipzig, benefit from the embargo. This created chaos and legal suits (including for Sertürner

Fig. 2.5 Joseph Louis
Gay-Lussac,
c.mid-nineteenth century.
Photo courtesy of Edgar
Fahs Smith Collection,
Kislak Center, University of
Pennsylvania

2.1.3   Gay-Lussac's Stock Tip for the Ages

After a long four years in the wake of the Battle of Leipzig which brought disar-
ray to the region, Sertürner persevered. Though he was mired in legal wrangling that
kept him from pursuing further experiments with what he called morphine, famously
named after the dream god Morpheus. The time away from his chemical investigations brought renewed strength to his former treatises on the importance of the new
drug, and for the first time he was receiving attention from other networks, even
ones outside of the Germanic States. These chemical heterotopias were labs full of
quizzical brows, but his second round of investigations had piqued the interest of
one of the most famous chemists in all of Europe who was going to be an important
player in the history of alkaloids. His name was Joseph Louis Gay-Lussac (Fig. 2.5)
[11].[9]

---

because he ended up losing that second pharmacy) that jeopardized their businesses well after the
Congress of Vienna, 1814–1815.

[9]Joseph Louis Gay-Lussac (1778–1850) along with his colleague, Pierre-Jean Robiquet
(1780–1840), made a formidable pair of French chemists (although Gay-Lussac was also a physi-
cist). Gay-Lussac formulated what became known as Gay-Lussac's Law, which stated that if the
mass and volume of a gas are held constant, then the gas pressure increases linearly as the temperature
rises. Robiquet laid the foundation through his work for identifying amino acids, the fundamental
building blocks of proteins. He did this through recognizing the first of them, asparagine, in 1806.
Then using the industry's adoption of dyes, he uncovered alizarin in 1826. Capping quite a career,
he advanced modern medicine with the identification of codeine in 1832, a drug of widespread use
with analgesic and antidiarrheal properties (which will be discussed in Chap. 4).

Sertürner had come a long way since his days at the court pharmacy in Paderborn. He had entered the field by publishing his findings in the prestigious *Annalen der Physik*. What he had discovered was the very first near overdose of morphine in history. Somehow, he convinced three colleagues, we do not know who, to take part in the first human test of an alkaloid. Each half hour he gave all of them (including himself) the equivalent of three times the maximum dosage in today's medical usage. How all of them did not overdose is an act of mercy, but it also had to do with the supposed emetic that Sertürner administered which brought them all back from the precipice [8]. The anecdote and supporting evidence must have stirred Gay-Lussac when he read that latest installment. His own work was on a parallel track to Sertürner. Working in France with the pharmacist Pierre-Jean Robiquet the duo recognized the need for a new category of chemicals that included alkaline substances of organic origin. Sertürner had provided the key to the codex that was morphine, and finally someone had acknowledged what he had started some thirteen years before [11].[10] It just so happened there was a Frenchman ready to give him plaudits.

The political situation at the end of the Napoleonic Era was tenuous at best. The tiger (France) had finally given way to the shark (England). The Congress of Vienna, headed by the crafty Prince Metternich, had attempted to preserve the Old Order, while looking to the future. Nationalism was forging new identities as states were also turning to science to become badges of progress and extensions of their technological savvy. France, restored after Napoleon's second exile was weak and manipulated by the diplomatic maneuverings of Talleyrand as he instituted a New Order. Prussia, the strongest entity of the Germanic States, was unable to unite Junkers (their elite military classes) with other provinces, which were particularly Catholic in the southern regions [12]. Science too was deeply divided both across and within national identities. Sertürner experienced this division firsthand because he was deeply injured by the fact that Gay-Lussac would recognize his achievements and not his own countrymen. He offered a strong rebuke of those that had given him so much by stating [11]:

> It is unfortunately too frequent a disease with us, that we direct our eyes more to the Gallic and British, than to the Germanic soil…added to this is the amazement and greedy hatred of the Germans for foreign products…they appreciate each other so deeply that they do not prefer the patriotism inherent in our sciences from abroad to others without equal value.

Though Sertürner had his pulse on the division that existed among the nations of Europe that would not brim over until 1914, he himself was prone to switching loyalties when during the 1820s a group of French chemists claimed that he had not discovered morphine [8]. This rebuke spawned another invective tirade against the French, which was only softened a decade later when they offered him the prestigious

---

[10]Gay-Lussac saw immediately the value of Sertürner's research because he immediately ordered a French translation in 1817. This reminds us of the important service provided by Madame Anne-Marie Lavoisier when she accurately translated her husband's work before his untimely death in 1794.

Prix Montyon and a handsome endowment of 2000 francs. Clearly national pride was more nuanced.[11]

What Sertürner discovered through his training in plant chemistry was nothing short of revolutionary. He did not stumble on it. Certainly, chemistry can have that quality, but not in this case. Training, opportunity, and the new scientific and chemical methods of the age assisted in one moving forward as an accomplished chemist. Even more impressive is that he had handed the baton of experimental pharmacology to the next generation of chemists and pharmacists; quite a triumph for an apothecary's assistant. Whether they were of French or German decent, it mattered none. A vast network was primed to advance medicine into a new generation of opiates. What occurred next was an impressive integration that would move the laboratory into a new phase of professional development as a chemical heterotopia.

The new chapter in organic chemistry began with the discovery of not just morphine, but a bevy of what became known as alkaloids. Joseph Louis Gay-Lussac's efforts along with his connections to French scientific networks and juried journals that published the latest findings concerning morphine spawned an outbreak of potential integration. As both the chemical and medical communities saw it, these new developments could revolutionize pain treatment because it offered precision. For the first time, the use of this term became commonplace, and now that salts could possibly turn into acids, the field was ripe for invention. Almost immediately the pharmacology programs throughout Western Europe began clinical research based on the effects of morphine on patients. The French physiologist, Francois Magendie published observations by 1821, and many, many others followed suit to the point that morphine entered numerous compendiums of pharmacopoeia during the 1820s [8, 11].[12] France now was stable enough politically as it entered the Restoration Period to draw upon its specialized knowledge in proximate organic analysis in brand new ways [12]. The professional ambitions of pharmacists at the École de Pharmacie, such as Pierre-Joseph Pelletier and Joseph-Bienaimé Caventou, formed the next generation of specialists that made isolating these new alkaloids their mission. By 1818, just a year after Sertürner published his work on morphine, two other massively important compounds, strychnine and quinine, were discovered. Like a giant national chess contest, not to be outdone, the German Carl Friedrich Wilhelm Meissner, the owner of the Löwen Apotheke in Halle, stamped the era with an important sobriquet when he defined these classes of compounds as *alkaloids* (Fig. 2.6) [11].

---

[11]Montyon Prizes (Prix Montyon) are a series of prizes awarded annually by the French Academy of Sciences and the Académie française. They were endowed by the French benefactor Baron de Montyon. It should be noted that Sertürner did receive honors from Germanic sources. He received a doctorate of philosophy from the University of Jena from none other than Goethe for his work on morphine.

[12]François Magendie (1783–1855) was a French physiologist who in 1816 published his *Précis élementaire de Physiologie*, which described an experiment first outlining the concept of empty calories. He said, "I took a dog of three years old, fat, and in good health, and put it to feed upon sugar alone…It expired the 32nd day of the experiment." Magendie is also known for his rivalry with the English scientist Sir Charles Bell over who discovered the differentiation between sensory

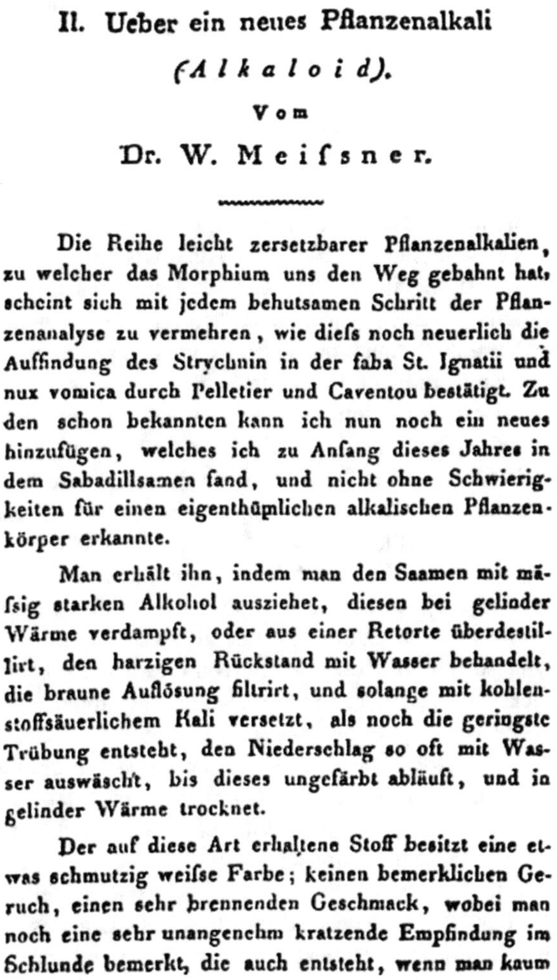

Fig. 2.6 The beginning of
the 1819 article by Carl
Friedrich Wilhelm Meissner,
where he defines the term,
Alkaloid. Photo Courtesy of
the Author

II. **Ueber ein neues Pflanzenalkali**

(A l k a l o i d).

V o m

Dr. W. M e i f s n e r.

Die Reihe leicht zersetzbarer Pflanzenalkalien,
zu welcher das Morphium uns den Weg gebahnt hat,
scheint sich mit jedem behutsamen Schritt der Pflan-
zenanalyse zu vermehren, wie diefs noch neuerlich die
Auffindung des Strychnin in der faba St. Ignatii und
nux vomica durch Pelletier und Caventou bestätigt. Zu
den schon bekannten kann ich nun noch ein neues
hinzufügen, welches ich zu Anfang dieses Jahres in
dem Sabadillsamen fand, und nicht ohne Schwierig-
keiten für einen eigenthümlichen alkalischen Pflanzen-
körper erkannte.

Man erhält ihn, indem man den Saamen mit mä-
fsig starken Alkohol ausziehet, diesen bei gelinder
Wärme verdampft, oder aus einer Retorte überdestil-
lirt, den harzigen Rückstand mit Wasser behandelt,
die braune Auflösung filtrirt, und solange mit kohlen-
stoffsäuerlichem Kali versetzt, als noch die geringste
Trübung entsteht, den Niederschlag so oft mit Wasser auswäscht, bis dieses ungefärbt abläuft, und in
gelinder Wärme trocknet.

Der auf diese Art erhaltene Stoff besitzt eine et-
was schmutzig weifse Farbe; keinen bemerklichen Ge-
ruch, einen sehr brennenden Geschmack, wobei man
noch eine sehr unangenehm kratzende Empfindung im
Schlunde bemerkt, die auch entsteht, wenn man kaum

2.1.4 The First -Oid's Makeup

What exactly was an alkaloid? As a new grouping of naturally occurring organic
compounds they were later found to have basic nitrogen as part of a functional
group. What chemists and pharmacists discovered beginning in the 1820s were that
they showed prominence in biological activity both in animals and especially, in
humans. These biologically active compounds that were responsible for the claims
of relief and cure of ailments were contained in a mixture of compounds as Sertürner
delineated. The physiological effects varied from plant to plant as do their structures.

---

and motor nerves in the spinal cord. His treatment of animals in experiments also drew the ire of
many, especially in Britain.

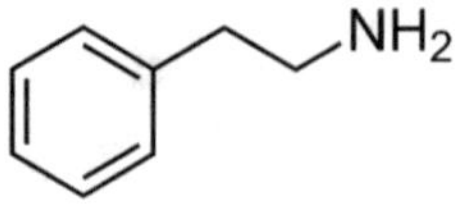

Fig. 2.7 β-phenylethylamine

Many alkaloids act on and disrupt the central nervous system through the subunit β-phenylethylamine (Fig. 2.7). Alkaloids act to mimic naturally occurring amines that are similar in structure, and serve as chemical messengers for the central nervous system. Chemists who also studied physiology uncovered that they are also partly responsible for our moods, and we would later come to know them as neurotransmitters. Known as neurotransmitters, these substances acted as a chemical bridge to nerve impulse transmissions between neurons. Common neurotransmitters we know today include epinephrine (adrenaline), norepinephrine, and dopamine to name a few [10].

In the end, it is the neurotransmitter that carries the nerve impulse from one to another. The neuron is made up of a large cell body called a soma, which is attached to a long stem-like projection termed as an axon. Filaments called synaptic terminals are attached to the end of the axon, and there are numerous short extensions named dendrites. These extend to the soma continuing through the axon to the synaptic terminals of one neuron to the dendrites of another neuron. In between that gap is the synapse. Within these friendly confines is the synaptic terminal of the axon where packets of neurotransmitter molecules are stored. An electrical current is generated by the exchange of positive and negative ions across the membranes of the neurons. These electrical currents flow from the soma and along the axon to the synaptic terminals triggering a release of neurotransmitter molecules that diffuse across the synapse and bind to the dendrites of an adjacent neuron. The dendrites contain receptors where the neurotransmitter binds to these receptors. Once the neurotransmitters bind to the receptor the message has been delivered to the receiving neuron. The receiving neuron then sends an electrical signal down to its axon and the whole process continues to the next neuron. Although no one could have predicted it, the first -oid to describe a chemical compound when paired with an alkaline, set off a chain reaction [8–11]. The chemical possibilities over the course of the next quarter century and beyond would yield investigations into some 30 new alkaloids that possessed a wide-range of effects on the human body. The age of the alkaloid was inaugurated.

2.1.5   The Valley of Chemicals

As the 1820s began to unfold, Europe experienced prosperity like never before, as it was an age where the alkaloid met industry. A new chemical heterotopia was upon them. For that to occur, it would take access to a deft combination of chemistry and practice, which for the time was not feasible. Of course, pharmacies had existed for centuries, first as apothecaries, but they were never scaled for distribution over wide

Fig. 2.8 Court Pharmacy, Berlin: Interior, c. Early to Mid-19th Century. Photo courtesy of Edgar Fahs Smith Collection, Kislak Center, University of Pennsylvania

expanses of terrain. Usually, they were bound to specific cities or regions. Now pharmacology was steadily becoming part of the curriculum at university laboratories. From a macro perspective, chemistry in general could revel in the fact that it was on the move as a discipline. Despite this shift, the pharmacies that pursued profit were where compounding took place because of their access to equipment and the skills necessary to produce product. Court pharmacies, which served principalities and were badges of an enlightened royal family continued to take the lead, especially in places like Berlin in the heart of the Germanic States (Fig. 2.8). Still, local pharmacies outside large cities built strong professional laboratories even though they could not produce uniform alkaloid products. However, those shops possessed access to the latest journals, were capitalistically-minded, and they had the skills, the ones that were progressive enough could build profitable ventures. A chemically-rich region like the southern Germanic states with easy access to a host of diverse elements would assist as well.

The first pharmacy to produce alkaloids on a mass scale was in operation consecutively since 1668 in the town of Darmstadt, Germany, just south of Frankfurt, which had recently become part of the new German Confederation. It was there that Heinrich Emanuel Merck (1794–1855), a direct descendant of company founder Friedrich Jacob Merck, first constructed the germ of what would become one of the titans of the global pharmaceutical industry. Merck's ancestor had purchased the Engel-Apotheke (which translates to Angel Pharmacy) in Darmstadt back in the seventeenth century.

Fig. 2.9 Merck Company Complex, Darmstadt, Germany, c. mid nineteenth century. Photo courtesy of the Merck Corporate Archives

This Merck, either because of guilt, zeal or a combination of both, studied pharmacy in Berlin and Vienna before returning to his father's Engel-Apotheke [13]. In 1816, though we are not sure how the workers felt, with his apprenticeship complete, he became the principal. His studies had exposed him to plant chemistry (like Sertürner before him) and he was occupied with research on the chemical construction of natural material. Like many of his Germanic competitors and French counterparts, he and his pharmacy team successfully isolated alkaloids and attempted to prepare them in a pure state. By 1827, Merck ramped up production and developed chemical processes that were streamlined for large batches. What he could accomplish, in time, made him the most important distributor of all the known alkaloids to other pharmacists, chemists, and physicians in the region. At the time of his death, some 50 workers were employed at his chemical-pharmaceutical factory, and upon his death, his three sons were poised to assume the business (Fig. 2.9) [14].[13]

How did Merck do it? Why did he succeed where others stumbled or in many cases never got off the ground? Certainly, the level of competition in the mid-1820s was at a fevered pitch to the point that pharmacies would gut one another to gain

---

[13]The pharmacy today is still part of Merck KGaA, but the original building was destroyed during the Second World War in the Allied bombing campaigns. It has since been rebuilt. Merck did not stop at pharmaceuticals; rather he diversified in 1838 by running a candle factory. He also became a leading citizen as a member of the town council of Darmstadt, and his family had deep connections to intellectuals like Goethe. And, if that was not enough, he served as a court consultant and worked on the well-known homicide case of the Mistress of Görlitz in 1850.

access to the secrets of alkaloids that were now available to anyone who could read and understand German, French or English. Merck found a way, but it was not just timeliness, industriousness or luck or chance or geography or history; no, it was all the above. Merck was poised to build an empire of chemicals because to begin with he was part of an apothecarian-pharmacological tradition that spanned not just decades but centuries. Since he was invested as a family member, he took full advantage of this past by apprenticing in the family business. Yet, he did not matriculate in his own backyard; instead, he went to large and diverse sets of chemical heterotopias which were central to German and even Austro-Hungarian industrial development. That knowledge of his family's past when paired with what he saw in those large court pharmacies became the basis for an overhaul in practice once he returned to the bustling metropole of Darmstadt in 1816 [15]. That date was significant because it was about the same timeframe when Gay-Lussac acknowledged and printed the alkaloid research of Sertürner. Within a decade, Merck built a progressive laboratory that was filled with the latest equipment and hired the best that the region in and around Frankfurt had to offer. One final piece of the Merck heterotopia though was significant—the types of earth around Darmstadt. Merck's family had the good fortune of being from one of the most chemically-rich regions in the world [15]. Like nineteenth century Britain or modern China, who both were and still are blessed with huge amounts of readily accessible coal deposits, the Merck chemistry laboratories were geographically in the middle of a veritable alkaloid dreamland [8, 11]. That spelled opportunity for someone like Heinrich Emanuel Merck, who possessed pluck, connections, good timing, knowledge, and history. These combined to assist him and his offspring in the development and production of morphine and a host of other alkaloids. An empire of chemicals was now at their disposal, and Europe was all too ready for the promise of pain relief.

Alkaloids, the unrealized dream of the Lavoisiers and their laboratory, had profound physiological effects on the human body. From a global perspective, the arrival of modern organic chemistry and pharmacology inaugurated a series of divisions between the university, the apothecary, and the pharmacy. The curtain rose on a new era as industry unfurled the flag in front of the production facility—a new heterotopia. As Sertürner's dog found out firsthand, some alkaloids could induce fatigue, create a sense of euphoria, elevate mood, or in extreme circumstances, depending on your neurotransmitters, a total collapse of your nervous system. The benefits of morphine and these other powerful drugs brought chemical hope to those seeking pain relief before 1860 (much as opioids would in 2018); and of course, Merck, and eventually their competitors, were all too ready to provide them. What nineteenth century societies in the West would come to understand was that when alkaloids were abused due to overuse, these agents could easily become addictive and have multiple side effects that ranged from constipation to weight loss. This was especially true in a place where healthcare professionals and pharmacists were fundamentally divided. Enter America.

2.2   Cooks as Chemists: America's Test Kitchen and the First Opioid Addiction Crisis

Only a woman can understand a woman's ills.
—Lydia Pinkham, c. 1875 [16]

2.2.1   Just What the Doctor Didn't Order

Texans are known for their health elixirs, like the prune juice based one from Dublin, Texas called Dr. Pepper. In the late 1880s, a raconteur from the *Lone Star State* by the name of William Radam came up with the ultimate medicinal drink; take water, a few drops of red wine, some acid, and bill it as a *Microbe Killer* (Fig. 2.10) [17]. Setting up seventeen factories, Radam exploited the discoveries in the laboratories of Pasteur and Koch by portraying the complicated diagnoses offered by doctors and health officials as superfluous. The real charlatans were the doctors, he said, and the age of patent medicines, that had nothing to do with the first portion of the name, gathered speed.

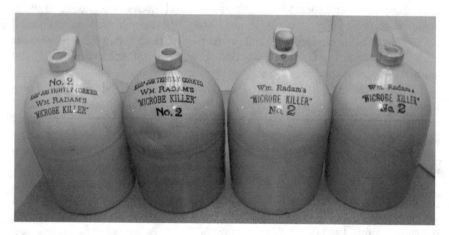

Fig. 2.10 Examples of William Radam's *Microbe Killer No. 2*, Crock Jugs, c. 1890. Photo courtesy of the author

2.2.2 Nostrum Republic

The first opioid addiction crisis in America did not begin in the early twenty-first century; rather it was in the late nineteenth century.[14] The origin was in a heterotopia you might have heard of, but would least expect—the kitchen. Cookery has a long-interconnected history with chemistry and vice versa [18]. Both possess instruments specifically designed for certain practices and employ those that are professionally-trained and those that have amateurs in their midst. As spaces, the kitchen and the laboratory heterotopias developed along similar lines and their platelets overlaid one another; by which ingredients, chemicals if you will, are placed in a certain order and reactions occur after a heat source is applied. Even the idea of wafting, using your hand to draw smells carefully towards your nose to judge quality so as not to be overwhelmed by fumes, finds a correlation. What is fashioned is a final product that is something wholly different than what came before. Thus, the laboratory was a kitchen and vice versa; and by the mid to late nineteenth century the Industrial Revolution would have a profound effect on them both [19].

As we saw previously, once the university laboratories of Europe took the lead, the professionalization of chemistry was underway. These chemical heterotopias imported raw opium and exported morphine by the boat-load, especially after 1860. From companies on the European mainland like Merck and British firms such as J. F. Macfarland, T. H. Smith, and T. Whiffen, the production of alkaloids that would treat everything from menstrual pains to dysentery flooded the markets [20]. In America, the process was much slower as science was still temporarily relegated to the pursuit of white gentlemen scientists. Before Alexander Bache and his cadre of Lazzaroni, or scientific beggars as they were affectionately known, helped to found the National Academy of Sciences, most experiments in chemistry were performed in private laboratories, especially when it came to pharmacological ventures [21].[15] Public colleges, which existed before 1860, and later, land grant colleges founded

---

[14]If you read most articles in print culture in 2017 most of them from *The Atlantic*, *New Yorker*, *New York Times*, and pretty much any other publication all place this crisis within the past 40 years.

[15]With the United States on the road to disunion in the 1850s, Bache would need to act quickly to peddle his idea of a national academy that would be devoted to science. To be sure there were a few examples of local and national organizations devoted to geology or the arts and sciences, but few were more than clubs in Bache's eyes. These entities were not concerned about serious scientific research nor were they focused on peer reviews, laboratory results, and international collaboration with other Western countries, like France, England, and the Germanic states, which were setting the standards for scientific inquiry. Bache watched as America plunged itself into regional battles over slavery and sectionalism. He began to gather a cadre of like-minded scientists to his cause, and cleverly dubbed them the Lazzaroni, after a group of beggars in Naples, Italy that sought shelter in a local hospital named St. Lazarus. As professional scientists, the Lazzaroni became the first scientific lobby group in Washington D.C. Made up of professionals they included physical scientists, who were close to Bache and his survey work, but they also included several others from disciplines like chemistry and biology. Wanting to develop an educational hub for science they spoke in meetings, published proceedings, and chased Congressmen down halls to sell the idea of a national academy. They found willing listeners since states were beginning to discuss the building of new universities and centers for learning, which in turn would feather their own nests if federal funding for these entities were secured. Bache was adamant that a national clearinghouse be established because the

for the most part after 1865, did not offer a Ph.D. in chemistry nor did they have adequate facilities. Most labs were built at private colleges, such as Yale, Princeton, Harvard, and Columbia and were the only places with chemistry professors at the time. Compared to their European counterparts they were poor imitations and lacked the most innovative equipment. If you wanted to study chemistry German was the place to be.

Though America was short on chemical contributions they did inaugurate what could be termed the first modern war. The American Civil War (1861-1865) accelerated technological developments, especially that which was associated with medicine. The modern hypodermic needle of the mid-nineteenth century became the preferred morphine delivery device and allowed those that knew how to find veins, more precision [22]. Milestones were reached on the British Isles beginning in 1844 when an Irish doctor named Francis Rynd reported the first recorded injection, in 1851, when Dr. Alexander Wood perfected the all-glass syringe, and finally, in 1858 when a London surgeon, Dr. Charles Hunter, supposedly stole Wood's thunder and dubbed the device, (based on the Greek words for *under the skin*) as hypodermic. The device was timely because as with many examples from warfare that converge in history, medical knowledge had grown by leaps and bounds. At first, surgery encampments, also called field hospitals, were scarier than the battlefield. Still, instruments and sterilization were becoming more refined and better understood, thus moving beyond previous practices. Like most events associated with extreme numbers of deaths in the modern age, soldiers were left to pick up the pieces of their shattered lives. Injected with morphine, when they could get it, forced habits saddled them with what became known famously as the *soldier's disease*. Combat would truly never be the same, and neither would those addicted in the twentieth and twenty-first centuries [23].[16]

Along with these new innovations and habits, advertising also became more prevalent and cheaper to produce. Printing firms began to use new colorization methods, and glass blowers literally cracked off glassware that would hold new chemicals for consumption. Pharmacies displayed the latest products. They, not the doctor's office, were the primary location in towns and cities where medical knowledge was disseminated. It would remain as such until the 1930s. Pharmacology in the late nineteenth century continued to be a trusted business, and was intimately linked to chemistry as we saw from their European counterparts [24]. Whereas doctors were perceived as charlatans, pharmacists, especially after 1820 in America knew chemistry. Compounding took creativity and a thorough knowledge of everything from tree bark to the chemical composition of alkaloids. Pharmacists, known as chemists or physics

---

Lazzaroni could govern by setting standards in the field, which in turn would eject those that did not conform to the latest and best practices. He was ruthless and quite unbending in this regard about what the country needed. Privately, he approached both Northern and Southern Congressmen by lobbying for science as a tool for defense and to build a powerful university system [21].

[16]The soldier's disease phrase is oft-used in the story of morphine. We do not mean to disparage it, but there is little quantifiable evidence to discern how morphine was distributed during the War and how many continued to use afterwards. Based on David Courtwright's work, this was a hidden epidemic. See footnote 13 for further commentary [28].

in Europe, could be dangerous. When they did not apply rigor to their craft, they could easily poison their customers. After the Civil War, they continued to gain in stature and strength within the medical communities of America. However, they were also swayed by profits, and those were made through the sale of patent medicines. The stage was set for a crisis, as Americans in the Gilded Age were bombarded by promises of pain relief like never before. A potent combination, as chemistry, pharmacology, and hope, albeit in fits and starts, coalesced around new products that were loaded with alcohol and opiates. The age of nostrums had begun [25].

2.2.3 Kiss the Cook

Lydia Pinkham was a nineteenth century chemist and pain-relief specialist. She was not a member of the American Pharmaceutical Association (APhA) and certainly her name would not be uttered in the same breath as Madame Curie. She was more like a cook-chemist, but her concoctions were certainly well-known once released after 1876. Pinkham was part of the patent medicine movement in the United States and probably one of its best-known provocateurs of what were termed nostrums [26]. After the Panic of 1873 destroyed her family's prospects, the former teacher and progressive fashioned her brand at the urging of one of her sons into *Lydia E. Pinkhams's Vegetable Compound*. Displaying her motherly face on her own advertisements she tapped into a specific niche, females who did not want to take their personal medical issues to a misogynistic profession (Fig. 2.11). Coupled with a choice investment and well-placed advertisement in the *Boston Globe* that cost the exorbitant sum of $60.00, she turned her bottle into a household name. What Pinkham and countless others proved was you did not have to be a professional chemist to sell medicine. What nostrums spawned though was the first opioid-like crisis in America because of what was in the medication. Like the opioids of the twenty-first century, patent medications, especially laudanum, wreaked havoc on society as Americans debated how to prosecute pain safely without forming habits that were detrimental to one's own health [27].

Patent medicine companies, like Pinkham's were a mainstay in America, albeit for a brief time, because people desired a new kind of medicine. Mostly, it was tied to the Industrial Revolution, which turned farmhands into wage laborers, and provided employment for new factories. Pinkham did not remain in her kitchen like other women during the period. She was a businesswoman, a chemist, and most of all a distributor, probably one of the first massively successful ones. Once she transferred her lab to a factory setting for increased production not only did she hire workers to mass produce her nostrum, but she also obtained the proper supply lines to acquire the necessary chemical compounds for her new laboratory. This new space within the factory needed measuring instruments, heat sources, and processes so wage earners could translate a specific formula into an order that would produce a compound. Once assembled, Pinkham's patent medicine was not that, patented. In point of fact, this was just a popular name assigned to these liquids because in previous decades the government had given protection for exclusivity of certain ingredients [29].

Fig. 2.11 Lydia E.
Pinkham's Vegetable
Compound, front side,
advertising card, c. 1888.
Photo courtesy of the Helen
Cushman Trade Card
Collection, Hagley Museum
and Library

Yours for Health
Lydia E. Pinkham

Patented or not, what Lydia Pinkham honed in her laboratory was not just ground herbs that were put into a labeled bottle. Rather she developed over time in her kitchen-lab a complex combination which relied on one of the staple principles in the history of chemistry, namely the systematic trial and error methodology. She may not have had a team like the French's Pasteur or the German's Koch behind her, but she had a network of important people in her Lynn, Massachusetts neighborhood. This consortium took part in a nineteenth century version of the medical trial, and they were tough customers that demanded therapies that worked. If the treatments did not pass muster with them, then it would not have a prayer to become a product that could be found in pharmacies across America. The compound she created that became so popular among her peers included a range of herbs and chemicals. A blend of true unicorn root (Aletris farinose) and pleurisy root (Asclepias tuberose) as active agents, it was probably the main ingredient that was so attractive to its paying customers—alcohol [28].[17] Although *Pinkham's Vegetable Compound* and other nostrums claimed that alcohol and also morphine were only used to chemically fortify, the root of the issue for the medical community was the compound produced

---

[17]There is no evidence to support the fact that Pinkham used opiates, but it should be mentioned that other nostrums did.

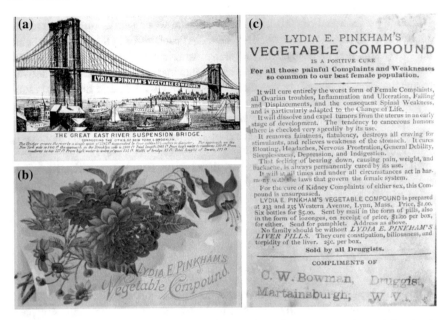

Fig. 2.12 **a, b, c** Trade cards, giving the impression of Pinkham's popularity and showing its national reach in accessibility, c. 1880. *Note* The reverse of the second trade card marked with the name of the druggist, C. W. Bowman, Martainsburgh (misspelled on the card), West Virginia, contacted directly by Pinkham's). **a** Photo courtesy of the U.S. Food and Drug Administration; **b, c** Photos courtesy of Nicole Girouard

in the lab had no listing of ingredients [29]. Thus, as temperance groups and other progressive societies, which pointed out the ills that patent medicines leveled against unsuspecting clients saw it, this did not establish the legal aspect of buyer beware.

Lydia Pinkham combated these claims against her company by doing something those chemistry labs and opioid producers have not done even to this day; she opened a chain of dialogue between her company and those concerned consumers. Entreating customers to write her directly, she set up lines of correspondence which opened a conduit directly to "Mrs. Pinkham" and her heterotopia. Women, who felt maligned and marginalized by male doctors, who purported to know best, responded in droves. They flooded the Lynn post with stories, of collapsed uteri and dizzy spells, advice for pain relief of migraines, and all in an attempt to steer clear of the doctor's unsterilized scalpels [29]. Pinkham's Front Office reported back with advice and kind words. Even though they were not necessarily medically sound or built up hopes, they still were taking what they scientifically and chemically believed, in turn, providing the best medical advice for the period. Like those with very little pocket money, wage earners took solace in the affordability. The company even guaranteed that no men would see the letters that were received, which instilled the idea of sanctity between a doctor and a patient. Even though Lydia Pinkham died in 1883, her mission to provide pain relief that was backed by chemistry continued (Fig. 2.12) [30, 31].

Though professional organizations (at least what was so at the time) such as the American Medical Association (AMA) and the APhA scoffed at the patent medicine trade, they still had to respect it. Too many Americans in the North, South, and even out West were sold on the positive treatments that nostrums claimed to deliver. Pharmacies stocked and restocked them because they sold, and in the 1880s alone *Pinkham's Vegetable Compound* cleared $300,000 in profit each year [26, 29]. Such numbers would make New York City's Boss Tweed envious, and for a woman to own a nostrum empire was truly ground-breaking. During the five years before her death, Lydia Pinkham had helped to redefine not only patent medicine, but her company expanded the definition of the chemistry laboratory. But, it was not all positive because pharmaceuticals, particularly patent medicines, were about to turn into an opioid-like crisis.

2.2.4 Eat Me, Drink Me

Before this crisis arrived, it is important to understand the state of the world after 1820. Global industrialization continued to accelerate at an increased pace after the Napoleonic Wars (The War of 1812 in America), and as we will see in Sect. 3.2, would have a profound effect on the definition of what constituted a chemistry laboratory. By 1860, global connections rated at greater speeds than they had in previous centuries. Land-based empires were monoliths compared to the ones that invested in new steam-powered sea travel. Transoceanic cables were being trolled off the backs of those ships, and it would not be long before a queen would speak to her subjects on the other side of the world. States that innovated by the 1870s, like Japan and Germany, would unify under a different model that emphasized a new kind of expansion and national essence [32]. Medically speaking, the expansion of chemistry and the professionalization of healthcare delivery were directly related to struggles over hegemonic power. The British-led Opium Wars beginning in 1839 had opened the Qing Dynasty to the West.[18] To obtain Chinese tea, British merchants used opium from India to obtain precious silver, which was then traded back to the Chinese in Canton for important teas [33]. As British property was destroyed and military action was pursued unequal treaties were levelled against the empire.

Opium smoking world-wide was perceived as lascivious behavior and recreational use was at an all-time high. Members of the American press portrayed the Chinese race as dangerous because dens for smoking immobilized its inhabitants. Opium was perceived, as it was during the British-Chinese Wars, as a means by which Asian men could lure white women into sexual situations where they could seek advantage. Most Americans of course did not require an opium den to *chase the dragon*, as it became

---

[18]It should be noted that the Chinese Emperor Yung Cheng prohibited the smoking of opium and its domestic sale (except by obtaining a license) as early as 1729. After the death of the powerful Qianlong Emperor, most nineteenth century rulers in China were unable to enforce the ban, despite the best efforts of their government officials (an example would be Lin Zexu).

known; rather, they could obtain a fix from the products produced by the numerous chemistry laboratories, just like the one Lydia Pinkham's company built. Nostrum producers continued to churn out new and improved products that contained tinctures laced with opiates. This was *the secret ingredient* in teething powders, analgesics, and of course, patent medicines [27].[19] Advertisements played upon people's hopes and fears, but because they had such a significant effect on their patients, the public tolerated abuse.

Rivaling these chemical heterotopias was the growing membership of both AMA and APhA. As a recent establishment, these organizations sought to wrest authority away from what was derided as quack medicine, and instead emphasized that professionally built healthcare delivery that was overseen by trained doctors and pharmacists was preferred. Particularly, the APhA took the lead by imposing standards, defining professional practices, lobbying state and local governments for better standards, and creating the United States Pharmacopoeia (USP) [34]. This regularly edited handbook was more than just a list of commonly used prescriptions coupled with savvy advice. With teeth, the USP was meant to establish scientific criteria, thus linking the latest developments in the chemistry laboratory with the people that prescribed and disseminated the drugs themselves. Since the late nineteenth and early twentieth centuries were the age prior to combination drugs (to be discussed in Sect. 3.2), for the most part only single entries and compounds were listed. Reflecting rather than legislating on what drugs were simply good or bad, the book did not have an immediate effect on the industry because most pharmaceuticals were not patented. That said, the APhA did see patent medicine as the bane of their existence, but their problem was one of quality control. Instead of forming an alliance with the medical profession, both sides subverted the other by prescribing and selling nostrums; a practice that continues to this day. One can conclude that the ready-made products, chock-full of alcohol and opiates, were just easier to prescribe and profit from, rather than steadily developing a nomenclature for just practices and good governance [35]. The definition of what constituted a laboratory could be extended then to include the pharmacies and doctors' offices that were integral to the prosecution of pain. Despite

---

[19]Individual patent medicines rose and fell in popularity over the course of the nineteenth century. The term quack was first applied during the Middle Ages to those that sold or *hawked* their wares by shouting or *quacking* in a loud voice on the street. Later, quacks were thought to be those that pretended to possess medical skills. Questionable diagnoses and claims of relief were all part of the quack patent medicines. In America, labels used all kinds of words or phrases to entice those that were suffering. Descriptions included: wonder worker, wizard, snake oil, nerve syrup, and sometimes were endorsed by both real and pretend doctors. The thought, from a marketing standpoint, was that this would lend creditability to the product, and snooker customers into thinking they were taking a quality product; the forerunner to the modern notion of a prescription. Once drug stores became more prevalent and started to serve meals at counters, many types of patent medicine companies began to convert to making what we know today as soft drinks. Examples include, Dr. Pepper, Coca-Cola, and others that became known as sodas, due to their carbonation, and were served in retail shops from fountains that in the future served ice cream. For two excellent collections that contain a host of patent medications include the Hagley Museum and Library in Wilmington, Delaware, https://www.hagley.org/research/digital-exhibits/patent-medicine, and the Smithsonian National Museum of American History in Washington D.C. http://americanhistory.si. edu/collections/object-groups/balm-of-america-patent-medicine-collection [25, 27].

a perceived professionalization campaign to link their two organizations, both the APhA and the AMA had a difficult task ahead of them because authority was not so uniform, nor simple, to claim.

2.2.5   Doctors Without Borders

The crisis surrounding opiate usage stemmed from the popularity of not just patent medicines, although that was hugely significant, but also from the laboratory's long-standing concoction called laudanum. *Harper's Magazine* reported that 300,000 lb of opium was being imported into the United States each year beginning right before the Civil War [20, 31]. [20] Whether these numbers were accurate or not is inconsequential. Rather the material point is that raw opium was present and being readily converted. As we will see in the next section also imported from Europe using alkaloid processes developed in the late eighteenth century to form opiates. These in turn were converted by all kinds of labs into what were called tinctures of opium. The primary opiate of the second half of the nineteenth century that was prescribed, sold, and marketed became laudanum and countless had what was termed a *habitué* [30].[21]

---

[20] Americans had attempted all sorts of opium importations schemes since the founding of the Republic. In fact, New Englanders in 1840 brought in roughly 24,000 lb of the stuff, which caught the attention of the United States Customs Bureau. They promptly placed a duty on the imports in order to attempt to curtail future shipments into the country. Later, *Harper's Magazine* actively covered the post-civil War opiate usage after 1865. Authors, including Horace B. Day who ghost wrote the Harper published, *The Opium Habit: With Suggestions as to a Remedy* (1868) went through numerous printings even after his death in 1870. His book added perspective by weighing in on how to cope with addictions. Most works just wanted to relate experiences that sold sensation rather than offering solutions to a major problem like addiction [36].

[21] It is difficult to ascertain how many people suffered under the yoke of opiate addiction since records were not kept with much accuracy and the term was not used until the twentieth century. The term habitué was used frequently to describe the habits of those that hooked. Again, for a fabulous regional study see [28]. What we do know is that states across America, especially out West were aggressive in their action against them. In 1872, California passed the first anti-opium measure, which held that "the administration of laudanum, an opium preparation, or any other narcotic to any person with the intent thereby to facilitate the commission of a felony." However, this initial attempt failed to control unlawful use of opium in the state. Connecticut, in 1874, became the first state whereby the "narcotic addict" was declared incompetent to attend to their own affairs. The law required that he be committed to a state insane asylum for "medical care and treatment" until he was "cured" of his "addiction." In 1881, the California legislature passed a law making it a misdemeanor to maintain a place where opium was sold, given away, or smoked. Only applying to commercial places, the law targeted the opium dens frequented by immigrant Chinese laborers. Smoking opium alone in a residence was not covered by the legislation. In the same year, California became the first state to construct a separate bureau to enforce narcotic laws, and one of the first states to treat addicts. During the last quarter of the nineteenth century, the western states continued to pass legislation restricting use of opium. Nevada's 1877 law was the first actually to prohibit opium smoking; this made it illegal to sell or dispense opium without a physician's prescription, and prohibited the maintenance of any place used for smoking or otherwise "illegally using" opium. Other western states soon had similar laws, with most legislation directed at outlawing opium smoking, rather than stopping use of other substances.

As previously discussed, this drug, which could be described as an opiate, had a long history in the West and went through numerous incarnations as cooks worked as chemists, even using the word recipes to describe their mixtures. What varied the most was the percentage of alcohol content, which hovered around 48–50%. Three in particular are worth noting because they each form a line of historical progression.

1. *Sydenham's Laudanum (dated from the 1660s): "According to the Paris Codex this is prepared as follows: opium, 2 oz; saffron, 1 oz; bruised cinnamon and bruised cloves, each 1 drachm; sherry wine, 1 pint. Mix and macerate for 15 days and filter. Twenty drops are equal to one grain of opium.* [37] [22]
2. *Dr. Alvin Wood Chase's Recipes, Laudanum: Best Turkey opium 1 oz., slice, and pour upon it boiling water 1 gill, and work it in a bowl or mortar until it is dissolved; then pour it into the bottle, and with alcohol of 70 percent proof 1/2 pt., rinse the dish, adding the alcohol to the preparation, shaking well, and in 24 h it will be ready for us. Dose—From 10 to 30 drops for adults, according to the strength of the patient, or severity of the pain. Thirty drops of this laudanum will be equal to one grain of opium. And this is a much better way to prepare it than putting the opium into alcohol, or any other spirits alone, for in that case much of the opium does not dissolve. See the remarks occurring after Godfrey's Cordial.* [38] [23]
3. *Tincture of Opium (Laudanum), USP, attributed to the United States Pharmacopeia (1863): Macerate 2½ ounces opium, in moderately fine powder in 1-pint water for 3 days, with frequent agitation. Add 1-pint alcohol, and macerate for 3 days longer. Percolate, and displace 2 pints tincture by adding dilute alcohol in the percolator.* [39]

Each of these recipes calls for the ability to chemically balance a mixture by first acquiring a list of ingredients and then by using everyday equipment to turn it into a chemical rendition. Dr. Chase's version entreated everyone from barbers to jewelry makers to use his practical information that would rationally treat a host of maladies from sleep deprivation to any and all inflammatory diseases. In the Preface to the Tenth Edition (his work was incredibly popular and went through numerous printings), he carefully explained how you can have your own home laboratory equipped with basic household items [38]. However, the construction of a kitchen-lab heterotopia required a warning that the *USP* did not. In the laudanum section under the

---

[22] By the nineteenth century, Sydenham's Laudanum (named after physician Thomas Sydenham, 1624–1689) was one of the longest running opiate compounds; which in some forms included sherry wine and a mixture of herbs. In 1676 Sydenham published his seminal work, *Medical Observations Concerning the History and Cure of Acute Diseases* in which he outlined the first early modern opium tincture. As mentioned, opium grew in several strategic locations throughout the Eurasian landmass. Silk roads and waterways dispensed it throughout the world, thus making it a truly international commodity. Interestingly enough, by most accounts it was the West of all places that abused opium and its derivatives [37].

[23] Dr. Alvin Wood Chase was originally from Cayuga, New York, and after attending medical school he established himself as a major force in self-help and in the dispensing of medical advice. He died in 1883 after a long life that was devoted to patients he never met in person.

heading for Godfrey's Cordial (which was a popular mixture of opium, treacle, water, and spices) in which he offered the following warning to his patients [38]:

> Then let it be remembered that the constant use of opium in any of its preparations on children, or adults, disturbs the nervous system, and establishes a nervous necessity for its continuation. Then use them only in severe pain, or extreme nervousness, laying them by again as soon as possible under the circumstances of the case.

What was Dr. Chase's point in offering such an editorial in the form of a medical diagnosis? Similar to Lydia Pinkham, Chase wanted to establish and then reestablish among his loyal readers that a portable laboratory could convert any kitchen or other space with a heat source into a veritable laboratory that could directly treat pain. By issuing subsequent editions he was taking advantage of his customer's desires for the latest information, and ascribing his name (and the limited medical training he had obtained over 30 years) to cures that if not properly understood could be detrimental to one's own health. In many ways, this was much more personal a touch than what the USP offered in its entry for laudanum, which only briefly mentioned the use of a percolator apparatus. The cold impersonal approach to healthcare delivery was in full swing as doctors attempted to establish their own authority against the thousands who would abuse opiates in the late nineteenth and early twentieth centuries. Up until 1914, one of several pent-up issues that encapsulated the era concerned how an assortment of doctors and prescription drug companies could challenge such a decentralized system while they agreed on so little.

How did the AMA and APhA arrest authority away from the Radams, the Pinkhams, the Chases or any other patent medicine company that stood in their way with snake oil as it was known?[24] In the wake of the crisis that touched every city and town in America in some form the interesting aspect of this from the outset was that laudanum was consumed by a diverse group from every aspect of the economic strata. Drug stores and the companies that supplied them continued into the 1890s to steadily increase their supplies of a variety of nostrums because that is what the public demanded. Conspicuous consumption, a phrase synonymous with the writings of Thorstein Veblen, ruled the market as Americans trusted in brands that contained heavy dosages of opiates [40]. With stores unwilling to pull their products as the year 1900 approached and an estimated 1 million users were addicted, how could an organization like the AMA establish and hold scientific authority? The late nineteenth century saw the crowning of a new chemical heterotopia that coalesced around the establishment of the modern industrial chemical laboratory. What they needed was a new kind of professional laboratory that drew on historical antecedents, while incorporating the recent lessons from the Industrial Revolution. Networked laboratory heterotopias like the ones associated with the development of a plethora of new drugs by trained industrial chemists might provide just the answer that was needed to solve the nostrum crisis.

---

[24]The term *snake oil* was a brand of nostrum, but it also became a moniker associated with products that made all kinds of fanciful claims. Thus, nostrums were dystopian because they heavily drugged their patients, thus providing only temporary relief from pain.

2.3 What the Networked Industrial Laboratories Produced: A Heroic Decision?

My duties are…to perform the familiar and dreary routine analysis of urine, blood, etc., for the clinicians. It is not for the advancement of their own understanding…but mainly as extra decoration for their clinical lectures…: a painted window on an artificial building.

—Max Pettenkoffer (Professor of Medical Chemistry, University of Munich) to Justus Liebig, discussing the direction of physiological chemistry, c. 1847 [41]

2.3.1 American Backwater

When it came to the chemistry laboratory as a professional space, the United States during the nineteenth century lagged far behind its European counterparts. It did have chemical companies that specifically manufactured alkaloids, like Lanman & Kemp & Kemp partnership in New York and Rosengarten & Sons in Philadelphia (both of these will be discussed in Sect. 3.1). Yet, unlike the nations of Western Europe that had a long tradition of pharmacy-based training programs, invested in rapid building programs for their own university laboratory structures, and engaged in over a century of competition that pitted one another in a war for chemical supremacy, Americans were outsiders. As already expressed they were unsure of what kind of laboratory they should build? One of professional chemistry—they were unsure of what that looked like. A pharmacy—they had those, but no tradition. To combat ineptitude, scientists after the Civil War booked tickets on the first transport they could obtain to discover how to duplicate their chemical betters. There were exceptions in North America though, those who tried to redefine their own laboratories by keeping pace in the hopes of professionalizing the field of chemistry. An American who best exhibited passion and wanted his students to understand the latest scientific innovations by having the opportunity to work in the best chemistry laboratory of the age was William H. Chandler of Lehigh University in Bethlehem, Pennsylvania [42].[25]

Chandler hired noted architect Addison Hutton who spearheaded the building (Fig. 2.13) of one of the most important architectural contributions to American chemistry in the nineteenth century (it was completed in 1885); a literal chemical dormitory was built for faculty, students, and the elements that had yet to be discovered [42].[26] Within the fireproofed structure (an innovation in urban environments) one could find large rooms that were divided into specific-use spaces, and equipped with ventilation units via a series of asphalt-coated chimney flues. Lab benches had

---

[25] William H. Chandler (1841–1906) was a professor, chairman, librarian, and acting president of Lehigh University. The building designed by Hutton is still in use today, but the Department of Chemistry has moved to another location on campus. The chimney-lined roof still forms a striking silhouette in the center of campus.

[26] Addison Hutton (1834–1916) was Quaker and a Philadelphia architect who was one of the principal designers of a wide-range of different building-types in Pennsylvania.

Fig. 2.13 Chandler Chemistry Building (with roofline of chimneys), Lehigh University, c. 1897. Photo courtesy of the Photograph Collection, Special Collections, Linderman Library, Lehigh University, Bethlehem, Pa.

room for 22 students which was an unheard-of amount for the time period. This particular contemporary modular bench layout possessed delivery services including gas, steam, vacuum, compressed air, and water. With a price tag of 200,000 dollars, Chandler had built a temple of quartzite, which would house the Chemistry Department for the next 90 years. Yet, while he added the tools for American students to become the heirs to the next generation of chemistry, an important link was severed. Temporarily, chemistry would not be an applied science allied to pharmacy or medicine [41]. Lehigh and its many fellow colleges and universities would have to wait for that connection to return. In the meantime, by the 1890 s, the laboratory practices of major European chemical firms, coupled with American distributors, would redefine the pharmaceutical industry and the dispersal of opiates, like morphine, as the twentieth century rolled on [43].

2.3.2 Liebig's Dream

The chemistry drug laboratory towards the end of the nineteenth century came in many forms and guises as a heterotopia. As we have already glimpsed you could find them in pharmacies, universities, within chemical factories and distributors, and even out back in kitchens that were attached and detached from the home. Staffed by a myriad of laypersons and professionals in Europe first, then as America followed suit, over time they were all increasingly influenced by the revolution that swept through industrialized nations and their burgeoning colonies. Alkaloids, like morphine, were incorporated into nostrums and created an addiction crisis as different types of labs pumped them out at will. Complex European networks of chemical production fueled by new global industries placed the laboratory at the center and with that shift, novel processes became integral to pharmaceutical development, especially in some parts of Germany. At midcentury, some laboratories were at least initially influenced by physiological chemistry which brought with it a focus on biochemistry and engaged the medical field in ways that honored the past and integrated the future [44]. Under the auspices of this new field, an organic umbrella promoted the chemistry with the best intentions, a science with a social conscience, if you will [45]. A chemical case that best exemplifies how applied sciences, like chemistry, medicine, agriculture, and pharmacology could produce an impactful product was the development of beef extract.

To this point, it is 1847 and you are a working-class member of a malnourished European society. Reaching levels of subsistence in the West is not easy, but wages are on the rise as the Industrial Revolution continues to globally increase with speed over the next twenty years. The eighteenth century political revolutions in America and France saw ideology turn radicalized and led to the rise of Napoleon as Master of Europe as well as a series of putsches and reform riots across the Atlantic World [46]. How important was the modern laboratory as a nineteenth century chemical heterotopia? For that matter, what role did it play as a link in the chain of industrialization and the formation of new synthetic opiates?

Before we get to those new drugs, we have to discuss their origins. Those lie in globalization, which was not germane to just the twentieth or twenty-first centuries. As a system it functioned at times like a well-oiled machine as ideas, peoples, technologies, and science cross-pollinated and churned to reform issues that before had cratered even the most-complex societies [47]. Synergy advanced industries that before were divided by oceans, land, and imaginations. But, there was inequality. Along with massive increases in industrialization, the West also experienced shifts in divisions of labor and widening gaps between rich and poor, which altered the patterns of disease and public health. The global scramble for resources led to continued colonial inequalities. To solve the problem of the high costs of European beef and in order to feed the working classes so they could run at peak efficiency a chemist, Justus von Liebig, walked down to his local German newspaper (Fig. 2.14). There he posted what would be the first of many want ads. Liebig was not peddling anything though; he was not a salesman, just a chemical revolutionary. He had only

Fig. 2.14 Justus von Liebig,
Engraving from an 1845
painting by Wilhelm
Trautschold, c. 1845–1863.
Photo courtesy of Edgar
Fahs Smith Collection,
Kislak Center, University of
Pennsylvania

one stipulation. Anyone could have his formula at no charge, but they just had to promise one thing: mass production had to be affordable to the consumers [48].[27]

2.3.3   Lab Ethnography

As Liebig made that famous trek to all those postal facilities, how was he able to produce something like a chemical composition for what would become beef extract? The chemistry laboratory by the 1840s had progressed since the days of Sertürner and his canine. No longer was the lab run as an individual enterprise, now it was a stage on which actors played parts of varying degrees. Those with experience were ranked higher than those with less talent. As director, Liebig conducted a complex chemistry of decidedly male egos (there was no Madame Lavoisier in this laboratory), emotions, and especially, how chemical and scientific knowledge was translated. A piece of evidence from 1842 (Fig. 2.15), when examined from an ethnographic and material culture perspective, reveals precisely what the laboratory was becoming. Namely, as the nineteenth century progressed, a dramatic space where action and reaction were experimented with just as much as the results from complex practices. By utilizing thick description, we see in the *actors* posed in situ [49, 50].

---

[27] Justus von Liebig (1803–1873) was the son of a pharmaceutical and chemical dealer. He was apprenticed to an apothecary, which, as discussed earlier, was before 1820 were loosely linked to the study of chemistry. Dissatisfied under this tutelage he moved to Paris to study under the chemist, Joseph Gay-Lussac. With the help of Alexander von Humboldt he joined the faculty at the University of Giessen, where he remained until he received the chair in chemistry at the University of Munich in 1852.

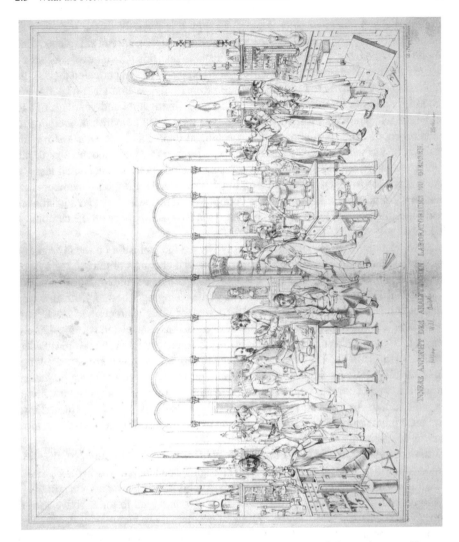

Fig. 2.15 View of Justus Liebig's Innere Ansicht des Analÿtischen Laboratoriums zu Giessen, c. 1842, for a closer view of this source see the link: http://sceti.library.upenn.edu/sceti/smith/scientist.cfm?PictureID=439&ScientistID=186 (Peter Morris in his excellent book, *The Matter Factory: A History of the Chemistry Laboratory*, uses this particular image, but offers little analysis of the source; other than commenting that he believes it to be staged or an idealized version of the laboratory [2].) Photo courtesy of Edgar Fahs Smith Collection, Kislak Center, University of Pennsylvania

At first, the scene might be interpreted as an idealized vision of Liebig's newly designed research laboratory, but instead if we examine it from a different set of points of view, we can uncover what transpired on a daily basis in this increasingly-professionalized chemical heterotopia.[28] Though this scene lacks the inclusion of the farsighted creationist of biochemistry, Justus von Liebig, we can still glean some important conclusions. What the image reflects is that this was an evolving space where the more advanced students carried out their research and perfected their skills in a laboratory constructed for the purpose of creativity—a first team of sorts. We see large windows that serve the dual purpose of illumination and ventilation, and oil lamps which allowed for work into the evenings. A built environment equipped tabletop room, storage cupboards for specialized apparatus, and even a stove for heating, which was designed to reflect the transference from the Neoclassical to the Empire Styles.[29] Liebig's charges in this scene include both advanced and less so, and the latter are marked by certain clothing that reflects their station—members like that on a junior varsity. These costumed pupils would reside in the older and smaller portion of the original laboratory that started out as a guard house on the campus of the University of Giessen [2, 48].

If we examine this drawing more deeply we see the inclusion on the far right of one of the leads who was one of the most influential chemists of the second half of the nineteenth century (Fig. 2.15). His name was August Wilhelm Hofmann. The son of Paul Hofmann, the architect of this game-changing laboratory, August was matriculating towards his Ph.D. in chemistry under Liebig's direction in 1841. Two years later, Hofmann became one of his most trusted assistants and was most likely one of Liebig's chief teaching assistants in the lab. Every chemical heterotopia needed one, and Hofmann had the chemical heritage to prove it. In this particular drawing, he is standing, possible imparting advice, another important communication device in the lab, to a classmate who looks to be heating a solution in a test tube with an alcohol burner. Advisement such as this was how textbook knowledge turned into chemical practice, and was a descendent from the apprenticeships that Sertürner took part in fifty years prior [51].

In the rendition, we also observe the bench design, which Hofmann's father oriented towards an open plan coupled with storage space that was readily accessible. The junior Hofmann could simultaneously watch his other classmates and still oversee a distillation experiment; thus, multi-tasking could lead to an effective use of time. While working with Liebig, Hofmann developed groundbreaking techniques in organic chemistry, some of which included obtaining aniline from coal tar. His work at Liebig's laboratory with organic bases would help develop a better understanding of natural alkaloids. Though Hofmann would eventually leave Liebig's laboratory within a few years, he carried a new chemical heritage from Giessen that would allow him to construct his own version of a modern laboratory wherever he

---

[28] Material Culture is an interdisciplinary approach that engages object-based analysis. The work of historians, curators, architectural historians, and those interested in how objects can tell important stories about the past have changed the way we look at subjects like the history of science. For an excellent appraisal of the field, see [49].

[29] The particular stove in the center-right portion of the rendering not only heats, but includes festoons and swag patterns which were reminiscent of the Neoclassical Style of the late eighteenth century. Empire elements are slowly being incorporated into the rest of the space, such as the size of the windows and their ornamentation in utilizing Corinthian column blocks (Fig. 2.15).

went. The physiological tree blossomed in London where he became the first director of the Royal College of Chemistry in 1845.

Other major actors include Adolph Friedrich Ludwig Strecker. He is positioned as the fourth person from the right (Fig. 2.15), with his back to us standing at a fume cupboard. Strecker was a student of Liebig's when in 1840 he earned a Ph.D. By 1846, he would become a private assistant for the principal. During his tenure, Strecker was known for working on reactions involving organic acids. In this scene, he was most likely working on some of those reactions. Organic acids can have unpleasant odors, so performing experiments within reach of the fume cupboard is logical. The chap sixth from the left, with his right index finger extended, (Fig. 2.15) was another prominent member of Liebig's cabinet of chemists by the name of Heinrich Will, who was in the lab at the master's request and would receive his Ph.D. in 1839. In this rendition, he appears to be advising another classmate about the piece of glassware that he is holding or perhaps he is seeking counsel on how to proceed with the next phase of his experiment [51].

The source not only depicts advanced students, but as mentioned, includes those who may be making the transition to more experienced ones. Modes of dress in both Europe and America underwent a massive shift from the eighteenth into the nineteenth century as wigs and breeches were officially jettisoned for men, and the corset became a thing of the past for well-positioned women. Fashion and medicine were intertwined in the laboratory as well as we see in Liebig's heterotopia. Before white lab coats, men (since women would not occupy positions there regularly until the twentieth century) were expected to dress in the 1830s and 1840s in much the same way as they did in public spaces. They wore top hats, long coats, trousers, and footwear that all resembled that which would be seen in parks, at the opera, and on the university campus [2]. Aprons of course could be donned on a case by case basis to protect a prized shirt or fashionable pair of new trousers, as they were called on the Continent and pants in America. Clearly, a laboratory hierarchy existed based on the several types of material culture examined in this image, which is best exemplified by the member of the staff with a basket of glassware. Carrying possible broken pieces that are in need of recycling, the individual depicted was certainly not an advanced student. In the foreground, there is also a person wearing a common apron and sitting at what looks like a large mortar and pestle (Fig. 2.15). He was probably a novice student working his way through the social and chemical hierarchy. Interestingly enough, he looks as though he could have benefited from a pair of safety glasses, since his left eye appears injured possibly by a projectile made of glass. Despite new regulations and formalities, the laboratory was still a place of danger. Architecture, hierarchy, costume, and safety, all combined to make this chemical heterotopia operate within a usable network, which in turn, would produce something that socially impacted lives, reflected pride in the nation, and would have a lasting impact on the future of the chemistry laboratory as the industrial revolutions marched on [52].

2.3.4   Chemical Nexus: A Framework

To contribute to society at large, what Liebig and his hierarchically garbed students unlocked, first at the University of Giessen and later at the University of Munich, was the key to what would become known as beef extract. Like morphine before, the process involved turning something from one entity into another. In this case, it was breaking down raw meat into gelatin. This liquid was produced in a factory and could be boiled into a beef broth which would not spoil or decay [48]. Just as Gay-Lusacc realized the promise of Sertürner's compound, Liebig needed a cohort that would assist with an industrial translation. He desired a catalyst who understood the gravity of his discovery, someone who appreciated his social and ideological bent, and most of all, knew where a bevy of cattle could be readily transported to Europe. He checked all of those off the list when he received a letter from one Georg Christian Giebert. Himself a scientist and engineer with connections to Belgium, Giebert happened to read about Liebig's offer while in Uruguay [53]. On a sojourn in country while inspecting rail lines, he noticed the dead carcasses of cows that never made it to market and were simply left along the ties. With the assistance of a British rancher named Richard Hughes who possessed thousands of cattle along the northern end of the Pampas, Giebert now had access to a slaughterhouse and a deep-water dock in Fray Bentos where vessels could port. After meeting Liebig in 1861, he needed one last piece to start the chemical locomotion needed to produce the extract—funding. What makes an international project of this magnitude get off the ground—credit. Luckily for Giebert his contacts in Belgium, where merchants for generations had carved out a financial structure that had competed with both the British East India Company and the Dutch VOC for financial supremacy, provided the help. The Belgian banks, which had deep connections with the region of South America where Hughes' ranches were located, were all too willing to invest in Giebert's venture because they knew a chemical opportunity when they saw one [54]. Huge numbers of wage earners deprived of the proteins needed for subsistence and growth? It was a chemical gold mine.

Yet by the end of 1862, Giebert's laboratory-factory was not as profitable as was previously conceived, even though he could produce the extract for less than one-third the cost than could be found in Europe. Like the IPO that is led by the millennial who has not quite done his homework, so it was with turning Liebig's formula into a veritable super food. What was the issue? Why wasn't this great idea working properly? The answer resides in the laboratory. It was not scaled properly. That is, the conversion rate that was necessary lacked the needed output to keep pace with demand. Making chemicals is an intricate process that requires not only funding, but technical expertise. Giebert's gambit did not adequately think about the chemical processes involved in such a venture. Certainly, he had assistants (in Antwerp) and chemists (such as a Mr. Seekamp in Fray Bentos) placed in strategic locations in between the beef in Uruguay and the distributors in the Germanic states [54]. Despite this network, he and his investors thought that transporting a lab more than 5000 nautical miles away through an intricate system of socio-economics would be possible. When viewed from a longer perspective such as this, it would not be an easy task. As future pharmaceutical companies would discover, turning complex chemical processes into consumer products was not so simple.

In 1866, a major shift in prospects for the company arrived the same year the Prussians destroyed the Austro-Hungarian army in a Bismarckian conflict that would lead to their unification of Germany after 1871. When the Dutch investors sold their shares to an English company that offered its shares in a limited-liability format, Giebert saw the opportunity to enhance the chemistry laboratory-factory in Uruguay. This move was crucial as capital flowed in that could enhance volume and production levels. The name of the corporation was changed to Liebig's Extract of Meat Company, thus accentuating the role of the dean of biochemistry while continuing to promote the liquid as a popular medical product produced by a professional chemistry laboratory [48, 53].[30]

How did Liebig change history chemically? The answer lies partly in the development of a chemical structure that he helped to construct through a series of experiments that were part science, part industry, part humanity, some good old-fashioned luck, but mostly due to a network of internal and external networks of development. His ability as a chemist to solve a problem such as hunger was limited only by his lack of connections both inside and outside the laboratory. Once chemical equations could be translated by the laboratory along the rail lines of South America, beef could be converted into extract. Yet, it took the translation by Giebert of Liebig's work and the link to Richard Hughes' beef from Uruguay to trigger the reaction. Nothing though could have occurred without Belgian and eventually British investment that connected to not only the working classes, but also the soldiers in armies across Europe that would also use the extract. In turn, that capital could not have happened without the laboratory.

The story of Justus von Liebig and his beef extract is an oft-told one by historians of the nineteenth century technology and those that study social integration during industrialization. Never has it been interpreted though through the lens of the field of chemistry and applied to the development of pharmaceuticals. What we see in this example is how the role of the laboratory as a flexible heterotopia could communicate and integrate complex physiological and chemical ideas, which could even be transported across oceans. Liebig's vision of an applied science would be imparted to his students, who were helping to build the chemistry departments at universities throughout Europe. Their influence on the soon-to-be-founded industrial laboratories of chemical firms would be essential. A new intersection of science, industry, medicine, society, and a connection to a pharmacological past brought much promise for the future of the field and for modernity. In the end, this chemical system, this symbiotic circle, would continue to function in many different capacities and does even to this day as major manufacturing firms produce opioids. Yet, somewhere along the line, the social component disappeared, replaced by concerns over the bottom line.

---

[30]Interestingly enough, historians often fall short in relating how by the turn of the century extracts were perceived as inefficient sources of protein, especially since the use of refrigeration was on the rise.

2.3.5   *The Learners Become the Masters*

The international system of industrialization fueled a set of revolutions that impacted wages, gender dynamics, and upward social mobility. By 1870, humans turned into not only engines of change, but unifiers, as the Germanic states, which had remained disparate for so long, now coalesced around factory production that would trump consumption. Chemical products in particular, as they had during the rise of Merck, became a marked success. Although the industry did not employ a huge number of workers, it was integral to building Otto von Bismarck's German machine.[31] After Liebig's dream began to fade, two branches of the chemical industry developed along different tracks [45]. The first emphasized heavy chemicals, which were alkali-producers of materials like soap and glass. Lighter chemicals encompassed a broad range of dyestuffs, which were used in the textile industry. Soon advances led to the discovery that these synthetics could be manipulated for use in perfumes, photographic materials, plastics, and most germane to this study, pharmaceutical products. As Heinrich Merck discovered in the 1820s, being geographically blessed had its benefits. In the region, raw materials such as rock salt (for sodium), potassium salts (for potash), iron pyrites (for sulfuric acid), and of course, coal tar (for those all-important, aniline dyes) were low-hanging fruit [55, 56]. The engines had all the fuel they would need.

It was at this point, roughly in 1865 as the American Civil War was winding down, where an important convergence occurred in chemical knowledge. With raw materials in copious amounts, new factories blossomed, and German chemists who had moved to France and Britain now returned to the Fatherland. They brought with them a vast set of experiences that were integral to chemical processes. Men like Heinrich Caro, Ludwig Mond, and Philipp Pauli along with former professors Carl Schorlemmer, William Dittmar, and probably most importantly, August Wilhelm von Hofmann formed an important foundation [56]. It was Hofmann (See Fig. 2.16, depicted in his private lab in Berlin, which is another type of heterotopia for those that had enough chemical clout), who had trained early on in his career in Liebig's heterotopia and was depicted in that 1842 drawing of the Giessen Lab. After he left for England and landed at the Royal College of Chemistry in London, he set to work on aniline dyes. With the knowledge he had gained back in Germany, and with his own innate ability, Hofmann himself developed blue and violet coal tar dyes [51]. Though he was lauded at industrial exhibitions, it was ironic that the person who had helped to develop these brilliant colors, along with cotton from places like Egypt and India, was part of something believed to be totally British. With the return of Hofmann and many other professionals, the chemical circle was now complete; when they left, they were but the learners, but now they became the masters; heirs to a chemical empire, yet, on the cusp of supposed greatness.

---

[31] Several popular histories of opium and opiates mention Bismarck as a habitual user of morphine. This particular type of scholarship spends the bulk of its time mentioning famous people from history that abused alkaloids. One reason for this might be taking an opportunity to make those from the past seem more human and flawed, anti-hagiography.

Fig. 2.16 A.W. von
Hofmann in seinem Berliner
Privat-Laboratorium, 1870
(This depiction of
Hofmann's private
laboratory is both functional
and ceremonial in nature.
Not only does it possess the
customary bench, with its
master of chemical
knowledge before it, but it
also displays busts of
influential chemists from the
past. Perhaps, Lavoisier,
Gay-Lussac, and Liebig were
included.) Photo courtesy of
Edgar Fahs Smith
Collection, Kislak Center,
University of Pennsylvania

A. W. VON HOFMANN
in seinem Berliner Privat-Laboratorium
1870.

With such vast chemical networks at their disposal, this cadre of chemists dispersed within a laboratory structure that was poised for success, especially in pharmaceuticals. Thus, unification of the state under Prussian command led to successful foreign wars against Denmark, the Austro-Hungarian Empire, and a major victory over its old nemesis, France. New opportunities for those who adhered to a Bismarckian World arrived and industry was directly alongside. The chemistry laboratory, a mainstay in university systems was transferred to the chemical firms that came to dominate the landscape. These included Friedrich Bayer & Company, Meister, Lucius & Brüning of Höchst-am-Main, the Badische Anilin-und Sodafabrik of Ludwigshafen, and Actien-Gesellschaft fur Aniin-Fabrikation (AGFA) of Berlin [56, 57].[32] It was in these types of firms where the physiological aimed laboratory merged

---

[32] Several scholars have argued that Bayer was an anomaly when it came to their production capabilities. While we consider this an important point, there were plenty of firms that worked towards solving a myriad of pharmacological issues of the day. Although, Duisberg's dream of a chemical conglomerate was not realized until the founding of IG Farben after the First World War, these firms still commanded an economic and political niche which was unparalleled even among the other industries of Germany.

with a reverence for a Merck-like apothecarian past in order to point in a new direction for drug production, which could enhance societies at-large. At least initially, profit was not an overriding concern during this golden age of pharmaceutical development as Liebig's dream of the triumph of physiological chemistry remained, albeit temporarily. Although slowly, what would wear this ideal away, would be competition from the other Western powers who attempted to keep pace with the German teams. The era, which would last until the eve of the First World War, would make the Second Reich the producer of three-quarters of the world's synthetic dyes, a heroic achievement; and the envy of those that had fallen behind.

2.3.6 Bayer Goes Public

How productive and influential were the German firms that built the most advanced laboratories in the world? As landmarks of expansion in aniline dyes such as alizarin, the coloring matter of the madder root coupled with Adolf Bayer's preparation of synthetic indigo, the stage was set for a pharmacological revolution [58]. It came as laboratory practices reached new heights of efficiency and efficacy. The personification of the new networked laboratory was the work of chemists like Carl Duisberg[33] and August Laubenheimer.[34] As university graduates they represented a different type of chemist. Both men were reared in environments that emphasized hierarchy, technique, research, and physiological chemistry. Each would have a major influence on German drug production before 1914. Duisberg presided over the powerful Bayer Co., which of course is credited with inventing aspirin, but they did more than that. Bayer, once Duisberg arrived as a chemist, moved towards the building of a series of laboratories devoted to research and training the next generation. Prior to his arrival most dye companies employed foremen who possessed degrees from places like the Krefeld Trade School. Duisberg was different. His academic background, devotion to separating physiological chemistry from material medica, verve for organization, and vision for a chemical industrial conglomerate expanded as he

---

[33]Carl Duisberg (1861–1935) grasped the mission that Friedrich Bayer envisioned. Like most Prussians, who took part in university training (he graduated with a doctorate from the University of Jena) and in military service, Duisberg combined savvy business sense (his first job at Bayer was as a dye consultant) with chemistry to be the organizing force behind Bayer at the turn-of-the-century. After taking a trip to America, he was greatly influenced by the design of vertical integration (two or more stages of production under one company) when he toured Standard Oil. For over four decades he presided over Bayer until it was assumed into the conglomerate I. G. Farben in the 1930s. For a thorough treatment of Duisberg's career see the article, [60].

[34]August Laubenheimer (1848–1904) studied chemistry in Giessen. He worked as an assistant and lecturer at the Chemical Institute of the University of Giessen and in 1876 attained the rank of associate professor. Later, he helped put Farbwerke Höchst on the map, by serving as a director and as a board member. During that tenure he wooed the future Nobel Laureate, Emil von Behring, to come and work for the company. By 1894, production began on a serum that could cure diphtheria and it was mass-produced as 75,000 vials netted over 700,000 marks in profit. Buried in the Old Cemetery in Giessen he was survived by son, the bacteriologist, Kurt Laubenheimer [61].

found new ways to closely link his own laboratory with the chemistry departments at universities like Berlin and Würzburg (which were shifting their own foci towards a new organic chemistry and away from the Liebig applied sciences approach) [59, 60]. Within two decades of his arrival at Bayer, Duisberg had helped to revolutionize the laboratory as a heterotopia that crossed academics with practice. In a similar vein, August Laubenheimer began as a scholar before joining the firm of Meinster, Lucius & Brüning where, like Duisberg, he stayed for over 30 years. Research in a new central laboratory led to a wealth of products including drugs that fought diphtheria, tetanus, dysentery, and foot-and-mouth disease [61].

The development of these chemical firms (by 1896 there were 108 of them with a capital of 332 million marks) had a set of unique checks and balances that allowed them to govern themselves and make key choices about their own chemical practices [62]. As long as they employed those that were committed to scientific pursuits and not profits, everything might work seamlessly. Bayer's development of a new drug is particularly integral to the development of opioids, and would present an interesting case in point as an early conundrum for the pharmaceutical firm. That drug was heroin. When Bayer first started to investigate the possibility of producing drugs, there was no formula for conducting research. In fact, Friedrich Bayer initially founded a Pinkham-esque laboratory in his home prior to acquiring the site at Leverkusen, which he purchased from a local chemist looking to get into the dye business in 1860. Really, it was not until 1896, when Arthur Eichengrün came on board, that the company began to enlarge its operations (Fig. 2.17). His job was to stimulate those chemical juices of ingenuity as a senior researcher, and unlike Hofmann who worked alone in his chemical advancements, now the research director would share in the profits of their investigations. Concurrently, hires were just part of the chemical expansion; additional spaces included chemical workspaces, but also offices, file rooms, and even libraries. Locks were placed on doors and hallways were constructed to guard access for those involved in certain aspects of development. Security was only in the initial stages of development, but the first application of research infrastructure had commenced [62].

2.3.7   A Chemical Rising Star

Alongside Eichengrün, Bayer added Heinrich Dreser, a chemist with a penchant for carpet bagging, and not one to conform to Liebig's alternative version of chemistry with a social conscience.[35] What Eichengrün lacked in ambition, Dreser made up for in spades. Conversely, Dreser did not possess the research sensibilities of Eichengrün. Thus, they formed a powerful team. Under Dreser's watch, the Bayer

---

[35] Heinrich Dreser (1860–1924) was born in Darmstadt, in the Valley of the Chemicals, also known as Merck's hometown. He was the son of a physics professor and received a doctorate from Heidelberg University before becoming a professor at Bonn University in 1893. He was known as one of the first directors of a major pharmaceutical firm to use animals on an industrial scale.

Farbenfabriken vorm. Friedr. Bayer & C° Leverkusen.
Aus dem Luftschiff „Zeppelin" aufgenommen
im Sommer 1913
vom Kommandeur des Luftschifferbataillons
Major v. Schulz, Cöln a/Rhein.

Fig. 2.17 Aerial view of the Leverkusen site looking west, taken from the airship *Zeppelin*, Summer 1913. Photo courtesy of the Corporate History & Archives, Bayer AG

pharmacological lab churned out new chemical products left and right. As the head of research at Bayer's Wuppertal laboratory, he led a unique organization that stood above all other firms. Bayer's success was due to the synergy they were able to harness not only with the personnel they acquired throughout the 1890s, but also through the blistering game of speed chess that they set for themselves with a system of trial and error. Starting with the work of Felix Hoffmann, who was guided by Eichengrün in 1897, the Bayer squad attempted to expound on previous inquiries into a new process by which to modify salicyclic acid [62]. Hoffmann's controversial discovery finally led Bayer to produce acetylsalicyclic acid (ASA) thereafter.[36] On the recommendation of his lead researcher, Dreser considered the potential of ASA, and summarily rejected it. Like so many discoveries in medicine over the course of

---

[36]The discovery of ASA was attributed to Hoffmann, but others within Bayer's ranks also claimed that they were the ones who uncovered it. The debate was not settled until the second half of the twentieth century when Hoffmann's name was finally attributed to his seminal work.

history, it would take generations to uncover the untapped potential of this *common* drug that became known as aspirin [63].[37]

Instead, Dreser turned his attention to another more potent one that was previously explored by the English chemist, C. R. Wright back in 1874. Ironically, Wright used a similar technique employed by the vegetable compound cooks, when he synthesized diacetylmorphine over his very own stove. Unlike with patent medicines in America that were given organic titles or named after their creators, like Pinkham, the white, odorless, bitter, crystalline powder was summarily given a technical name. In Germany, during the final two decades of the nineteenth century, advertising and patenting drugs were practices frowned upon and were against the law. Companies chose innocuous names and rarely toyed with the idea of marketing. All of that though was about to change forever.

2.3.8 Dreser's Dystopian Dream

Dreser was really the first chemist inside the bowels of the laboratory to first see the potential for diacetylmorphine as a non-habit-forming version of morphine—an -oid. As previously discussed, morphine use had become heavily stigmatized since the American Civil War, and so laboratories naturally sought any and all opportunities to replace it. As previously mentioned, healthcare delivery was desperate for professionalization as laudanum reigned and because of this opening, Dreser was all in. He knew about Wright's previous work, but pretended that it did not exist. As the freezing temperatures of winter gave way to a new spring in 1898, Dreser set his team to work. He gathered as many test subjects as he could. They fished sticklebacks from rivers, gigged frogs from local markets, cornered rats with humane traps, and bred rabbits that seemed to multiply easily before their eyes. Even Bayer's staff received complementary samples in what must have been the first time an industrial laboratory sponsored a *first is free campaign* or to use the term from the street, the pass out. Employees and lab technicians supposedly reported that it made them feel *heroisch* (German for heroic), and the appellation was spot on [64].

The rats who were smacked with that initial taste of the new drug returned again, and again, and again, to the very spot in their cages where they were first given the stuff. It was too enticing, and it took over their existence. We are not sure what became of those members of the Bayer staff that served as guinea pigs, but we can surmise that a similar fate met them. Like morphine, which over the course of the nineteenth century induced not only euphoric reactions and a numbing feeling, there were side effects that ranged from constipation to weight loss. Heroin was no different, except it created a much greater set of reactions and side effects. From a chemical standpoint, from 1899 to this day, whether heroin continues to be produced in a controlled environment of a pharmaceutical laboratory or in the chaotic environment of a makeshift, clandestine lab, the first step toward synthesizing heroin remains the

---

[37]For a chemical and historical survey of the history of aspirin see [63].

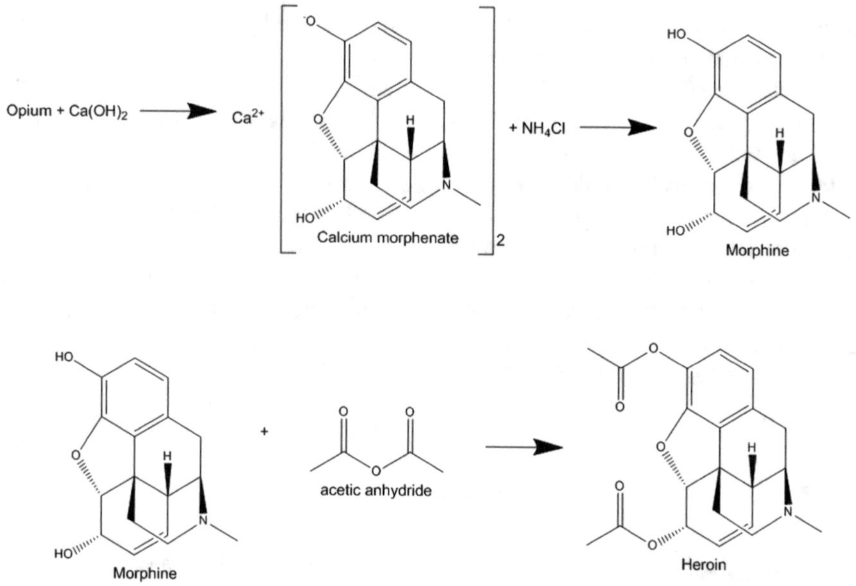

Fig. 2.18 The Heroin synthesis

same. As we related in Sect. 3.1, morphine was isolated from the opium extract of the poppy plant. In its raw state it was added to boiling water dissolving the opium into a clear brown liquid solution. After this was filtered of all the solid particles, calcium hydroxide or any compound containing elevated levels of lime content was added to the solution (Fig. 2.18). During this process that Bayer now scaled for larger batches for the first time in history, the lime was then converted from morphine into calcium morphenate. The insoluble impurities were precipitated out and the solution was then filtered again. The filtered solution was reheated and ammonium chloride was added to the calcium morphenate solution to adjust the alkalinity to a pH of 8–9. The solution was then allowed to cool. Within one or two hours, the morphine base was ready to be filtered and the resulting morphine compound could now be converted to heroin. Thus, synthesizing heroin from morphine continues to be a relatively straightforward process that requires only a few steps to complete. Perhaps, this speaks to the power of scaling and testing that Bayer commanded in their own laboratories? The main conversion step was acetylation of the hydroxyl (–OH) groups of the morphine molecule using acetic anhydride. Anhydrides react with the –OH group of a molecule to form an ester and acetic acid. After the heroin was produced, a series of recrystallizations and purification steps were performed. HCl could then be added to the heroin to yield the hydrochloride salt of heroin in order to boost absorption into the bloodstream [62, 65].

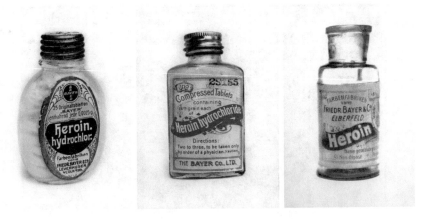

Fig. 2.19 Heroin bottles, Western markets, c. 1899–1920s Photo courtesy of the author

2.3.9 Follow the White Rabbit

Once this method was completed, what ended up being four times more powerful than morphine, heroin was presented to the Congress of German Naturalists and Physicians later that year. Bayer argued that it was more effective than codeine (ten times so), and of course, they made the claim that would be heard all the way to the twenty-first century, *it is not habit-forming* (see Fig. 2.19 for three incarnations of heroin for the Western markets); a panacea, for a new age and new century. Why was something as common as a cough the first malady named? In the nineteenth century whooping cough (known clinically as pertussis), pneumonia, and tuberculosis were rampant. As airborne diseases, that spread easily through coughs and sneezes, spread from class to class, fits were so debilitating death could be eminent. The company with a fully industrialized heterotopia who could legitimately claim to offer a cure for those ailments would be lauded as nothing short of heroic.

By 1899, Bayer's new product was flying. Where Liebig's beef extract had initially failed at scaling, the Leverkusen network was primed for production. With a Barnam-esque swagger of *there's no such thing as bad publicity*, Bayer banked on the restoration of much needed sleep through the quieting of those raspy and hacking coughs (Fig. 2.20). Thus, heroin's groundwork was laid by Dreser's version of a chemical heterotopia. He had turned the chemistry laboratory campus at Bayer into a networked set of spaces that could effectively generate, produce, and test new forms of pharmaceuticals [62]. Before releasing samples to the medical community, he had reached out to pharmacists who controlled those local journals and thus those vital links to customers. Free samples in the form of mailshots were dispersed throughout the West. Tagged with the iconic *lion and globe*, Bayer turned out literally a ton of

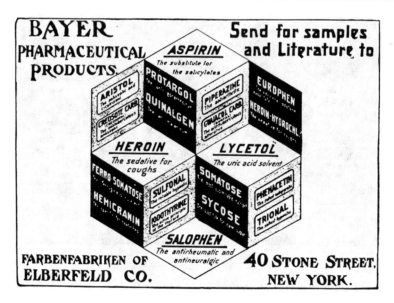

Fig. 2.20 *Sedative for coughs* advertisement by Bayer, building block format, *New York Medical Journal*, c. 1900. Photo courtesy of http://web.missouri.edu/~sherk/Bayer-heroin.jpg. Accessed September 25, 2017

the product that last year of the century.[38] It was a prophetic end to a century that had begun with Sertürner's tests on his dog; now the beginning of the new one seemed full of so much chemical hope. Of course, Bayer had the assistance of the German state behind them; it provided the most effective patent law in Europe, a favorable tariff structure, and support for scientific education and research. As the industry matured, it had protected itself against predatory practices. Little did they know that the demon would come from within.

Although it was exported to some 23 countries, heroin found a home in America. Morphine exports from Europe had reached new heights in the 1890s as distributors in major American port cities stocked their warehouses with a host of derivatives (See Sect. 3.1 for further discussion of this). With so many morphine enthusiasts already on board, the American Medical Association actually approved the use of heroin in 1906 (the same year as the Pure Food and Drug Act) despite knowing the possible risks associated with prolonged use [66]. Something else was occurring with its arrival across the Atlantic; now, it was being mixed with other liquids that manufacturers could market to consumers or could be disseminated as the first twentieth century street drug. Heroin, cheaper to manufacture and more potent than morphine, came in many forms, including lozenges, tablets, elixirs, and in water-soluble salts. Within a decade, Bayer had started to receive reports from drug-infested hospitals full of patients and members of the medical community decrying that Bayer's product was

---

[38]The symbol of the *lion and globe* was adopted later by the well-known brand of Burmese heroin called *Double Globe*, which coincidently has a similar package.

a runaway train. Ironically, on the eve of the First World War just before humans morphed themselves into weapons of mass destruction, Bayer pulled Dreser's wonder drug. They simply had no choice. Carl Duisberg, an advocate for the founding of a chemical conglomerate that could protect the industry much like the American Standard Oil, determined that aspirin was a much better asset to promote, especially since it sold itself [60].

2.3.10 Gateway to a Synthetic Age

What had the late nineteenth century laboratory begat—more diverse practices, more complex hierarchies, and more integrated chemists? The lab as a chemical heterotopia came in many different forms as it attempted to professionalize over the course of the era. Physiological studies had spawned biochemistry as an evolving set of practices in organic chemistry. Liebig's dream of a socially conscious laboratory and therefore a new integrated science had not quite held against the corporate structure that his students' generation had constructed. As the world hurtled towards an end to the long peace, nations in the West possessed such power in the synthetic materials they produced that the age of the nostrum seemed doomed since professionalization had secured a toehold. But, was this the case? Had the kitchen lab of the cooks been defeated by the high ethics and the progressive studies of chemistry and medicine that could be found in the elite institutions of Europe and America? It seemed as such. Yet, though the laboratory was the harbinger of a better tomorrow chemically, it had also produced heroin, a dystopian drug and one that would find its way to the street. The days of Sertürner and Gay-Lussac collaborating from a distance by actively communicating about a new alkaloid had given way to the era of Merck. Though Justus Liebig's attempts to revolutionize the laboratory and his influence over many of his students like Hofmann remained in principle, changes in the way biochemistry was defined was underway. The networked industrial laboratory, despite its best intentions, was beginning to be separated from its past connections to the applied science. And even more concerning than this, the major firms began to be more and more focused on the search for profits. Without consulting other fields, they forged ahead with drug manufacturing, thus inaugurating another step on the road of how an opiate became an opioid. Now the synthetic opened a new chemical door that would unlock a host of possibilities and consequences.

After Heinrich Dreser began to test his great creation he utilized animals, the staff at Bayer, but also in a page from Robert Louis Stevenson's *Strange Case of Dr. Jekyl and Mr. Hyde*, he also, like Sertürner, tested it on himself. The result was infinitely more destructive, since what he was ingesting was more powerful and certainly not heroic. By 1914, at the age of 53, Dreser was a wealthy man, so much so that he accepted an unsalaried position at a pharmacological institute that he founded in Dusseldorf. He lived out his days in Zurich where he died of a supposed cerebral stroke in 1924. Heinrich Dreser was integral to the development of heroin and introducing it to the world, but it also turns out, he was a junkie. Clearly this proved that the synthetic opiate had arrived, and anthropomorphism did not discriminate based on economic or social boundaries.[39]

References

1. American Chemical Society International Historic Chemical Landmarks (1999) Antoine-Laurent Lavoisier: the chemical revolution, Paris, France, June 8, 1999. http://www.acs.org/content/acs/en/education/whatischemistry/landmarks/lavoisier.html. Accessed 17 Sept 2017
2. Morris P (2015) The matter factory: a history of the chemistry laboratory. Reaktion, London, pp 47–49, 93–96
3. Donovan A (1993) Antoine Lavoisier: science, administration, and revolution. Cambridge University Press, Cambridge
4. Roberts W (1992) Jacques-Louis David, revolutionary artist: art, politics, and the French Revolution. University of North Carolina Press, Chapel Hill
5. Eagle C, Sloan J (1998) Marie Anne Paulze Lavoisier: the mother of modern chemistry. Chem Educ 3:1–18. http://www.springerlink.com/content/x14v35m5n8822v42/fulltext.pdf. Accessed 17 Sept 2017
6. Eddy M, Mauskopf S, Newman W (eds) (2015) Chemical knowledge in the early modern world. University of Chicago Press, Chicago
7. Cooper A (2007) Inventing the indigenous: local knowledge and natural history in early modern Europe. Cambridge University Press, Cambridge
8. Meyer K (2004) Dem morphin auf der spur. Pharmazeutischen Zeitung (in German) GOVI-Verlag. https://www.pharmazeutische-zeitung.de/index.php?id=26551. Accessed 15 Sept 2017
9. Booth M (1998) Opium: a history. Simon & Schuster, London
10. Bentley KW (1954) The chemistry of the morphine alkaloids. Clarendon Press, Oxford
11. Lesch R (1981) Conceptual change in an empirical science: the discovery of the first alkaloids. Hist Stud Phys Sci. 11:305–328
12. Vick B (2014) The congress of Vienna: power and politics after Napoleon. Harvard University Press, Cambridge
13. Rosen W (2017) Miracle cure: the creation of antibiotics and the birth of modern medicine. Viking Press, New York, p 158
14. Hawthorne F (2003) The Merck druggernaut: the inside story of a pharmaceutical giant. Wiley, Hoboken, pp 19–25
15. Ebner F (1968) Merck and Darmstadt: in perspective over the generations. Merck, Darmstadt
16. Danna S (2015) Lydia Pinkham: the face that launched a thousand ads. Rowan & Littlefield, New York
17. Starr P (1982) The social transformation of American medicine: the rise of a sovereign profession and the making of a vast industry. Basic Books, New York, pp 122–123
18. Freedman P (2007) Food: the history of taste. University California Press, Berkeley
19. Laudan R (2015) Cuisine and empire: cooking in world history. University of California Press Berkeley
20. Berridge V, Edwards G (1981) Opium and the people: opiate use in nineteenth-century England. Allen Lane, London
21. Jansen A (2011) Alexander Dallas Bache: building the American nation through science and education in the nineteenth century. Ger, Campus Verlag, Frankfurt

---

[39]Dreser remarried after the death of his first wife during the First World War. Historians have suspected that instead of taking an aspirin each day, which he was integral in marketing as well, he took heroin. This period is where the term junkie was coined after some who collects metal for sale in order to support a heroin addiction.

22. Hacker B (ed) (2016) Astride two worlds: technology and the American Civil War. Smithsonian Inst Press, Washington DC
23. Courtwright D (2002) Forces of habit: drugs and the making of the modern world. Harvard University Press, Cambridge
24. Estes JW (1986) Public pharmacology: modes of action of nineteenth-century 'patent' medicines. Medical Heritage 2:218–228
25. Young JH (1967) The medical messiahs: a social history of health quackery in twentieth-century America. Princeton University Press, Princeton
26. Higby G, Stroud E (eds) (2005) American pharmacy (1852–2002), a collection of historical essays. American Institute of Pharmacy, Madison, WI
27. Boyle E (2013) Quack medicine: a history of combating health fraud in twentieth-century America. Praeger, Santa Barbara, CA
28. Courtwright D (1983) The hidden epidemic: opiate addiction and cocaine use in the South, 1860–1920. J South Hist 49:57–72
29. Stage S (1979) Female complaints: Lydia Pinkham and the business of women's medicine. WW Norton & Co, New York
30. Courtwright D (2001) Dark paradise: a history of opiate addiction in America. Harvard University Press, Cambridge
31. Cowan R (1985) More work for mother: the ironies of household technology from the open hearth to the microwave. Basic Books, New York
32. Landes D (2003) The unbound prometheus: technological change and industrial development in western Europe from 1750 to the present. Cambridge University Press, Cambridge
33. Ellis M, Coulton R, Mauger M (2015) Empire of tea: the Asian leaf that conquered the world. Reaktion Books, London
34. McTavish J (2004) Pain and profits: the history of the headache and its remedies in America. Rutgers University, New Brunswick
35. Gabriel J (2014) Medical monopoly: intellectual property rights and the origins of the modern pharmaceutical industry. University of Chicago Press, Chicago, pp 107, 109, 224
36. Day H (1868) The opium habit: with suggestions as to a remedy. Harper and Bros Publishers, New York
37. Sydenham T (1848) Medical observations concerning the history and cure of acute diseases. In The Works of Thomas Sydenham, M. D. Latham R. G (trans), 2 vols. The Sydenham Society, London 1:29–268
38. Chase AW (1870) Dr. Chase's recipes, Tenth Ed. R. A. Beal, Ann Arbor, MI, pp 133–134
39. Dick WB (1890) Encyclopedia of practical receipts and processes. Fifth Ed. Dick & Fitzgerald Publishers, New York, pp 416, 447, 472
40. Veblen T (1994) The theory of the leisure class: an economic study of institutions. Penguin Twentieth-Century Classics. Intro R Lekachman. Penguin Books, New York
41. Kisskalt K (1948) Max von Pettenkoffer. Wissenschaftlichen Verlag, Stutgart
42. Billinger RD (1939) The Chandler influence in American chemistry. J Chem Ed 16:253–257
43. Adas M (1991) Machines as the measure of men: science, technology, and ideologies of western dominance Ithaca: Cornell University Press
44. Rosenberg C (1979) The therapeutic revolution: medicine, meaning and social change in nineteenth century America. In: Vogel M, Rosenberg C (eds) The therapeutic revolution: essays in the social history of American medicine. Univ Penn Press, Philadelphia, pp 3–25
45. Kohler R (1982) From medical chemistry to biochemistry: the making of a biomedical discipline Cambridge: Cambridge University Press, pp 9–19
46. Rabinbach A (1992) The human motor energy fatigue and the origins of modernity. University California Press, Berkeley
47. Rosenberg C (1992) Explaining epidemics and other studies in the history of medicine. Cambridge University Press, Cambridge
48. Brock W (2002) Justus von Liebig: the chemical gatekeeper. Cambridge, New York
49. Gerritsen A, Riello G (2015) Writing material culture history. Bloomsbury Academic, London

50. Egbert M (1991) Scientists' orientation to an experimental apparatus in their interaction in a chemistry lab. Issues Appl Linguist 2(2):269–300
51. Hofmann AW (1866) The chemical laboratories in course of erection in the Universities of Bonn and Berlin. W. Clowes and Sons, London, p 8
52. Day C (2017) Consumptive chic: a history of beauty, fashion, and disease. Bloomsbury Academic, London
53. Perren R (2006) Taste trade and technology the development of the international meat industry since 1840. Ashgate Publishing, Burlington, VT
54. Bell S (2015) New frontiers and natural resources in southern south America, c. 1820–1870: examples from northwest European mercantile enterprise. In: Trading environments: frontiers, commercial knowledge an environmental transformation, 1750–1990. Winder G and Dix A (eds) Routledge, New York, pp 47–68
55. Travis A (1990) Perkin's mauve: ancestor of the organic chemical industry. Technol Cult 31:51–80
56. Travis A (1993) The rainbow makers: the origins of the synthetic dyestuff industry in western Europe. Lehigh Univ Press, Bethlehem, PA, pp 35–44
57. Beer J (1959) The emergence of the German dye industry. University Illinois Press, Urbana, Ill
58. Meyer-Thurow G (1982) The industrialization of invention: a case study from the German chemical industry. Isis 73:363–381
59. Travis A, Hornix W, Bud R, Homburg E (1992) The emergence of research laboratories in the dyestuffs industry, 1870–1900. Brit J Hist Sci 25(1):91–111
60. Stock A (1935) Carl Dusiberg: 29.9.1861–19.3.1935. Ber Dtsch Chem Ges 68(11):A111–A148
61. Galambos L (1997) Networks of innovation: vaccine development at Merck, Sharp and Dohme, and Mulford, 1895–1995. Cambridge University Press, Cambridge
62. Wimmer W (1998) Innovation in the German pharmaceutical industry, 1880 to 1920. In: Homburg E et al (eds) The chemical industry in Europe, 1850–1914: industrial growth, pollution, and professionalization. Kluwer Academic Publishers, Dordrecht, pp 281–292
63. Rooney S, Campbell JN (2017) How aspirin entered our medicine cabinet. Springer Publishing, Heidelberg
64. Lenoir T (1988) A magic bullet: research for profit and the growth of knowledge in Germany around 1900. Minerva 26:66–88
65. Fernandez H, Libby T (2011) Heroin: its history, pharmacology, and treatment, 2nd edn. Hazelden Center City, MN
66. Liebenau J (1987) Medical science and medical industry: the formation of the American pharmaceutical industry. Johns Hopkins University Press, Baltimore

Chapter 3
Part Two: Synthetic Opiate Heterotopias

3.1 The Age of Synthetics

Heroin does not nauseate, and can be given in teaspoon dose as often as every two hours to adults, this dose of course being graduated in children according to the age, although they tolerate it where opium would produce untoward results…it can be given indefinitely without the patient turning against it.

—Dr. J. Leffingwell Hatch in *The Canadian Journal of Medicine and Surgery*, Volume XIV, No. 4, October 1903 [1]

There is but one safeguard in the use of these remedies; to regard them as one would regard opium, and to employ them only with the consent of a physician who understands their true nature.

—Samuel Hopkins Adams, reporter for *Collier's*, c. 1905 [2]

3.1.1 A New Laboratory of Progress

In 1900, citizens in Western societies exhibited a combination of zeal and practiced apathy towards the impending modernity of their worlds. Some were brimming with promise and hope, but most looked out their residences with trepidation. Wage earners had made strides, gaining access to consumer goods, and new types of commercial empires rose. However, storm clouds were on the horizon, as old alliances crumbled and businesses reacted in accordance. Europe within two decades would not be the same again, as Russia imploded then rose, France and Britain heaved to, Germany though not invaded was crushingly humiliated, and the United States emerged as a reluctant global player in international relations. Marxism now had an application, or so it was thought, and the stair-stepped progress argument for industrialized nations pointing towards a utopian society was reborn. As Modernism's promise came crashing down in the World Wars, for those that reveled in their own magnificence, science was a lever by which nations expressed themselves technologically; all the while crafting a unified monoculture that they believed was just as important to their own identities.

Chemically and medically-speaking, the previous century brought some relief to a plethora of pain, due to the historical connections between compounding pharmacies

© The Author(s) 2018
J. N. Campbell and S. M. Rooney, *A Time-Release History of the Opioid Epidemic*,
SpringerBriefs in History of Chemistry, https://doi.org/10.1007/978-3-319-91788-7_3

and their patients. However, like one of the side effects of morphine (constipation), with the advent of organic chemistry and the rise of industrial pharmacology, treatments continued to be blocked by a lack of centralized authority that could consistently control the deluge of the latest, though untested, materia medica. Drugs were still only recently dispersed with any measure of dependability during the 1890s with the majority being self-purchased, not prescribed. As far as quality controls for the time, some forms of a product might be stronger or weaker in reaction by the body. The factories were scaled properly as corporate capitalism gained a foothold, but there was still inconsistency and proverbial imitators hawking things that looked like professionally-made pharmaceuticals, but were not. The problem also existed of knowing what combinations of drugs would interfere with one another. Medical journals created a semblance of bridging many gaps, as the chemical community and healthcare professionals formed an unholy alliance. In the meantime, pharmaceutical production was about to be revolutionized, unbound from lines on a map, and expanded even further as companies took advantage of new forms of prescribing, such as fixed dosage combinations. Thus, universities, industries, small town mom and pop pharmacies, and eventually the federal government, all vied for the claim that their chemical knowledge of synthetic alkaloids was superior. Nostrums, previously cloaked by suspicious ingredients, were on the run as the curtain was jerked back, despite their ability to span socio-economic strata. As we have discussed, the evolution of the modern opioid began with the isolation of the first alkaloid—morphine, which promised numbness, but also led to side effects and addictive behavior. Meanwhile, the powerful chemical conglomerates of Germany turned their backs on physiological chemistry and instead university systems within the country separated departments that had previously, in Liebig's day and before, been intertwined with pharmacy and medical training. Now ensconced in their ivory towers of information across Europe and North America, the teaching of organic chemistry turned out a cadre of chemists that flooded those powerful dye companies (i.e. AGFA, BASF, Bayer, and Hoechst). By 1900, the focus of production began to shift to the full-scale industrial manufacture of pharmaceuticals. The road to the modern opioid hurtled forward in a vain attempt to end pain [3].

Devoted to making modern medicine because of the wealth of profits, the industrial laboratories, were now under the firm control of men like Carl Duisberg of Bayer on both sides of the Atlantic and Georg Merck with a major firm ensconced on American soil in 1891. These entities teamed with another form of a laboratory heterotopia that would once again redefine the notion of what we think this space was and is; enter the large-scale distributor. Like the forerunners to the modern McKesson, Cardinal Health, and AmerisourceBergen of today, these *labs* literally handled and at times mixed the alkaloids.[1] Not only that, but these companies possessed access to

---

[1]McKesson has a long history in America that dates back to the 1830s. After two different mergers over the course of the nineteenth and twentieth centuries they have built a powerful empire as a drug distributor that dwarfs all others. Currently in 2018, they are the 5th largest company in the country and gross over 200 billion dollars a year. They have become a major distributor of opioids and disseminate over a hundred million pills a week, which has drawn the attention in several media-driven pieces, namely by the news magazine *60 Minutes*, which has attempted to expose its

powerful lanes of commerce to create an interconnected network of drug creation that could be transported anywhere on the planet [4]. Not only did they handle logistics, this understudied extension of pharmaceutical production and sales represented the brands that they were channeling to the stores and then to the public. These specific laboratories that produced alkaloids required well-stocked equipment as essential, along with personnel that were trained in the most technical practices of the age, but were they ready to challenge past practices with new ones? How would the new laboratories of the twentieth century, which felt so confident in their abilities to master medicine, develop and distribute synthetic opiates that became opioids? The answer lies in the development of important allies who could engineer imports and deliver those products and pieces of equipment directly to doctors and healthcare providers who, as we will see, were themselves going through their own phase of professionalization. Into this complex world of drug design stepped governments that were attempting, through their ever-expanding agencies, to address society's ills by safeguarding the public. Pushed by muckraking journalists, therapeutic reforms were underway in what historians would come to call the Progressive Era. Comprised of all sorts of interests, medicine would see the transference of practices and knowledge from the pharmacy to the doctor's office and hospital by the 1930s. Through two global conflicts, the rise of authoritarianism, and the incorporation of new drugs, the pharmaceutical laboratory entered a new stage of rapid chemical development by the 1950s with the realization of what was nothing short of an empire, known as Big Pharma, and at their disposal, a set of powerful new synthetic alkaloids [5].

3.1.2 The Social Drug Network

At the dawn of the twentieth century, the world of drug design and manufacturing across the West was becoming a complex hash of networks within networks. Bridging aqueous territories from the Caribbean across the Atlantic, and eventually through the Panama Canal to the Pacific, was an important extension of the chemical networks [6]. The role of the chemistry laboratory and the development of processes that were prototypes fifty years prior, found even on the banks of the Rio de La Plata in Uruguay, were now commonplace, as oceans were traversed and land masses bridged by canals and rail lines. Alkaloids, like morphine, were now scaled and designed as never before in places that were not thought of before. The synthetic age was like a crusty brown piece of Pennsylvania scrapple that contained parts of parts; savory, but not necessarily good for you. Those wage earners with their pocket money were consuming distrustfully. Concurrently, doctors were hungry for the latest information on new drugs and distributors wanted to become speedy millionaires. Previously isolated, university professors were becoming half instructors and half consultants. Finally, there were those commercial chemistry laboratories that came in many different forms, who were desperately trying to keep up with the demands

---

corrupting influence on the pharmaceutical trade. That is quite a chemical heterotopia. https://www.cbsnews.com/videos/60-minutes-washington-post-investigate-deas-biggest-opioid-case/.

from their front offices. Healthcare slowly developed from a social good at the local level to an avaricious market where profiteering was paramount. Dye companies in Europe and in America knew that the key to establishing themselves was the ability to secure a patent.[2] More and more, these served as a legal means of suppressing those chemical cooks of the world and the rogue distributors who pirated their formulas. Patenting, before a frowned upon practice, would help to assist them in ending the nostrum crisis, and more importantly, helped in developing new products for emerging markets. The Age of Imperialism arrived just in time for Americans and ushered in opportunity in Latin and South America [7].

At this point, alkaloids, whether morphine, heroin, or as another derivative, became just one of many diverse products that drug manufacturers distributed. Disseminators of these products were not necessarily analogous. You could be a distributor, but not make alkaloids onsite in your warehouse. However, you could be a manufacturer, and if you had warehouse space and a dock near your onsite laboratory, distribute. The parent company, to use a term from the second half of the twentieth century, relied on distributors to represent their interests and also mete out their products. A long-standing company that engaged heavily in the drug distribution business illustrates this point well because it was internationally connected and built a reputation of fair dealing. Its name was Lanman & Kemp (established in 1808 and incorporated in 1920), and it was a multi-generational family firm of wholesale drug distributors in New York City. Their records document operations that run the gambit of the business in the years before, during, and after the development of modern pharmaceuticals [8].[3]

The drug business for them began in the 1840s during the days when Alexander Bache only dreamed of a scientific empire in North America. Founded initially by the Murrays, a prominent Quaker family in New York, they built their merchant house by selling a variety of materia medica throughout the Atlantic World. The warehouse included a laboratory heterotopia that produced what was known as *Florida Water*, a nostrum that was promoted as a cosmetic and for its restorative powers. With colorful

---

[2] As previously mentioned, doctors in the early years of the United States objected as much to the secrecy of patent medicines as they did to the actual drugs sold by salesmen as branded *potions* promising miracle cures (some of which worked, as we saw with *Pinkham's Vegetable Compound*). That same secrecy interfered with physicians' ability to dispense medical assistance. In Great Britain, patents were granted by the crown, which assigned medicines and protected knowledge. In the United States, the secret formulas and their direct promotion to the public under branded names disturbed healthcare providers. After the American Civil War and near the end of the nineteenth century, standards, trademark law, and even the expression, *patent*, evolved in meaning through the work of Francis Stewart, the firm, Parke-Davis, and through court cases such as Bayer's patent of aspirin (see Rooney S, Campbell JN (2017) *How aspirin entered our medicine cabinet*. Springer Briefs in the History of Chemistry. Springer, Heidelberg, pp 22–27) that made acetylsalicylic acid therapeutically. Opinion slowly came to accept patents as a means of recording the ingredients of remedies (more on this in Sect. 3.3).

[3] The Hagley Library in Wilmington, Delaware holds a significant cache of these files. They were acquired after a stamp collector pilfered all the international stamps from the envelopes. As for the firm, it was incorporated as Lanman & Kemp, Inc. in 1920 and moved to suburban New Jersey in 1957. Several of Lanman & Kemp's traditional products are still produced and marketed by Lanman & Kemp-Barclay & Co., Inc. in 2006.

labeling, it was an all-purpose product that was quite popular.[4] Towards midcentury, Murray added David Lanman as a partner who was also connected to the gentry, albeit from nearby Connecticut.[5] Upon Murray's death, the business continued under Lanman. At this point, the firm added opiates (several forms of alkaloids), medicinal and culinary herbs (always a connection between the former and latter, especially when it came to cookery), medical apparatus in the form of glassware, and a host of items including books and periodicals. Lanman made an interesting business decision in 1853 when he partnered with a man called George Kemp and his five siblings from Ireland. The Kemp team had lived the American Dream when their widowed mother brought the group to New York City where the streets were barely paved and the Five Points were a hotbed of all sorts of diversions. It was a major decision for the business because right before the Civil War it became D.T. Lanman & Kemp, and the timing was impeccable considering the impending crisis ahead [8].

As part of the wholesale drug trade a firm such as this profited from two major political and economic developments. First, was the advent (as discussed in Sect. 2. 2) of the American Civil War. Morphine needs were on the uptick due to the extreme numbers of deaths in makeshift butcher shops called field hospitals. Thus, firms like Lanman & Kemp were well-placed to provide the United States Quartermasters with the medical supplies they required (including hypodermic needles). This was a foremost contract and gave a massive economic boost, while fueling soldiers' recovery and the habits of addiction. The other aspect that was improved upon was the trade and connections that came afterwards. As industrialization ramped up even further to new Promethean heights in the 1870s and 1880s, Lanman and his sons cultivated business in Latin and South America, which meant navigating a different language and taking part in a range of local and regional cultural milieus. They travelled to these potential customers with the assistance of guides, a need for being able to communicate at times in the language of chemistry, and made connections through an aqueous network that was brimming with opportunity and danger. Throughout the late nineteenth century correspondence reached warehouses and clients in an inter-connected system (see Fig. 3.1). The mark of a company was quite literally the development of their own letterhead which served as a flag of their own empire. Lanman & Kemp wanted their symbol to be an edifice which housed their departments and laboratory for commerce [9]. Thus, new chemical heterotopias swung open for firms like this one that acted as a chemistry lab, warehouse, distributor, logistics manager, customs broker, travel agent, import-exporter, direct mail order provider, purchasing agent, and last but not least, commercial empire builder.

On the import side, in particular, Lanman & Kemp, backed by London banking firms, brought in huge amounts of raw material by paying the duties, which meant one needed the brokerage expertise, so as not to alarm the Customs House. This material originated in places as far away as Southeast Asia and the Mediterranean

---

[4]Murray & Lanman Florida Water Cologne New York is sold to this day in major chains across the World.

[5]David Trumbull Lanman (1802–1866) had deep connections to the American past as he was a descendant of both Governor Jonathan Trumbull and the artist John Trumbull.

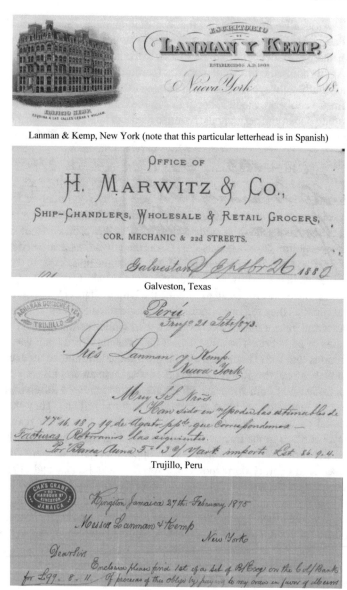

Lanman & Kemp, New York (note that this particular letterhead is in Spanish)

Galveston, Texas

Trujillo, Peru

Kingston, Jamaica

Fig. 3.1 Four examples of letterhead to/from Lanman & Kemp, c. 1880s. Photos courtesy of the Hagley Museum and Library [8]

region. In particular, Turkish opium was a major import. Though different from Indian opium, it was used by the company in their own nostrums, sold to a variety of outlets throughout America, and even re-exported out of the country. Lanman required a firm grip on the huge amount of correspondence (some was sent via telegraph in code in order to protect information and clients' identity), the numbers, and understanding of currency exchange rates in order to keep a complicated set of books.[6] The cooks needed their ingredients, especially in the South where opiate use was strongest, yet hidden [10]. Thus, a complex network, similar to the one that organized the system behind Liebig's extract, emerged to supply all sorts of materia medica up and down the socio-economic ladder with various and sundry choices. Despite the trappings of rampant runaway capitalism, businesses like Lanman & Kemp were about to receive some major competition once the future pharmaceutical giants Bayer and Merck arrived in America.

As discussed, the German juggernaut of dyes and drugs were taking the world by storm as factories turned out it seemed something new each day. However, even the big firms needed those all-important distributors who knew their networks best and could negotiate contractually in the interests of the home office. This was a tricky business, especially if the distributor was in another city or especially in another country on the other side of the Atlantic. This was the case with Bayer and Merck as they rose to prominence during the 1890s until 1914. They relied on the Liebig-esque system, but as they solved the scaling issue, they harnessed the chemistry laboratory by combining industrial know-how with university-led research into organics that became the most important cogs in this intricate system. Bayer and Merck had conquered supply chain organization with the rapid acquisition of raw materials, and the application of factory streamlining (similar to what Henry Ford adapted at his Highland Park Plant at Ford Motor Company). So, they were poised for expansion into new markets.

There was one main issue out of several that pertains to this arrangement. What if the distributor was not on the level? Simultaneously, cooked the product and the books? What if they sought their own advantage? Altered products? Used the home office's name improperly? All those scenarios happened to Merck, and in a different way, affected Bayer's heroin in the United States. What they did about it changed the pharmaceutical industry in America forever. As a company, we already know Merck had a long history of alkaloid development. They single-handedly revolutionized the market in Europe, but in the United States they relied on a distributor, such as Lanman & Kemp, to sell and distribute their products. Late in the nineteenth century, Merck employed a firm called Lehn & Fink (an unfortunate name considering what was afoot) as their American distributor. All was going well, that was until it was discovered that the distributor was taking E. Merck labels and placing them on a variety of non-Merck products. The home office was less than pleased because customers would easily think that the Merck name was associated with shoddy materials. In this business, your name was all you had, and Merck's history would not allow such a soiling.

---

[6]The company must have encountered interesting exchange rates when local mercantile outfits traded for nostrums and a plethora of other goods, by offering a host of items including produce, hides, and coffee.

Fig. 3.2 Advertisement in *Canadian Journal of Medicine and Surgery* c. 1903 (left); Merck Delivery Truck, c. 1912 (right). Photos courtesy of Merck

To remedy the situation Georg Merck, the son of Heinrich, dispatched a trusted and chemically connected member of the extended family to assume control. Theodore Welker arrived in America to a complex chemical heterotopia with the sole job of arresting control of supply lines and making the connections necessary to reestablish the Merck name. The home office used an old, a really old, model that had worked before in Darmstadt all those years before. They built an impressive pharmacy with cornices, stylized columns, and polished floors that served as a flagship headquarters in the center of the most bustling metropole on the planet. Even though he was forced to close the compounding pharmacy two years later from pressure from New York druggists that felt that this was unfair competition, their morphine production in their industrial chemistry laboratories increased tenfold. Thus, Merck & Company became a household entity with annual sales well over a million dollars a year in 1897 (it would become incorporated in 1908 and exceed three million in sales by 1910), and supply lines that reached as far as St. Louis, Missouri and into Canada [11]. Though still related to the original company back in Germany, they were now wholly separate. Merck would need that distinction when 1918 came around (Fig. 3.2).

Similar to their competition, Bayer also came to America at roughly the same time. Under Carl Duisberg's leadership and with products like Aspirin and Heroin (notice the capitalization since they were proprietary before 1914), Bayer was poised to dominate the American markets. Yet, it was not going to be easy. In 1903, just as George Merck was becoming an American citizen (adding the *e* to his first name for assimilation's sake), Bayer selected a new site where they built one of the largest and most up-to-date factories of its time. Known as the Rensselaer Plant, it became the American home for the production of both Aspirin and Heroin. Bayer was now extending its production lines, in order to subvert the Lanman & Kemp type competition by not only coming to America with their products, but also extending their influence into Latin American markets through advertisement (Fig. 3.3) and the hiring of Bayer agents.

Fig. 3.3 Advertisement in *Canadian Journal of Medicine and Surgery* c. 1903 (left); Spanish language advertisement for Bayer Heroin c. 1909 (right). Photo courtesy of the Corporate History & Archives, Bayer AG

Heroin in particular became a major seller, especially, as discussed in Sect. 2.2 for coughs and tuberculosis [12]. Ready markets also took the product and incorporated it into other mixtures that formed a veritable nostrum underground trading network, a new form of a future drug trade that will be discussed more extensively in Chap. 4.

Besides this sort of competition there were other forms of heroin that were developed in American laboratories. Possibly the most well-known was called Glyco-Heroin, produced by the Martin H. Smith Company in New York. Not too far from the Lanman & Kemp's proto-lab and distribution center, this concoction of heroin that Smith produced in its own laboratory combined heroin with sugar syrup. No records were found that Bayer sued Smith over patent violations; Smith was purchasing heroin first from Bayer, then converting it by adding their own ingredients. Billed in advertisements (Fig. 3.4) as aiding *therapeutic qualities*, Glyco-Heroin was sold as a treatment for coughs, bronchitis, asthma, laryngitis, pneumonia, and whooping cough. Glyco, as it was known for short, was a synthetic alkaloid and was priced to directly compete with Bayer. In fact, Bayer was fairly concerned that this new form of heroin would eclipse theirs. In order to stem this tide, the Rensselaer Plant went into full production after 1903. The rush to build new factories in North America by both Merck and Bayer reflect the desire to eliminate the middle man of the alkaloid trade from Europe. Now, with laboratories entrenched in the United States, they could defend their products both in person and on paper [11]. This allowed for legal action against those that rivaled their own chemist and ensured that their good names would be protected. As a first step to establishing authority over their competitors, pharmaceutical companies once again redefined the laboratory heterotopia and codified a means by which to control a complicated drug social network. Lanman & Kemp would survive of course, but now large industrial firms, the origins of Big Pharma, harnessed the written word like never before in order to shore up their influence.

3.1.3 How They Made Heroin Sound Healthy

The *Sears & Roebuck Catalog*, which was in circulation for almost 100 years before it was digitized, is not usually thought to be or even remotely associated with a chemistry laboratory. However, what is undisputed is that American consumers looked at this publication as an authority that combined taste, fashion, and choice during the late nineteenth and for most of the twentieth century. What may not be fully realized about the bulky page-turner was around the 1890s to roughly 1910, Sears, besides selling everything from brasseries to whole house kits, sold morphine directly from the catalog [13]. Not only could you purchase it, but you could also buy the syringes with which to inject yourself. Sears was just like any other company taking advantage of wage earners and meeting demand for new goods. Of the alkaloid derivatives they sold most were listed under the *Prescription Department* (this meant family recipes, rather than those prescribed by pharmacies) or sold in the *Patent and Proprietary Medicine* section of the catalog. They had enough currency with consumers and were so popular they could sell ice to Eskimos, as the saying goes. What Sears proved though was that print culture was pervasive, authoritative, and could manipulate markets and people by placing advertisements in a complete compendium. If you needed a house, done; how about the latest fashion, zip, to your door; and, as far as pain relief from the preeminent alkaloid of the nineteenth century, sign and press here.

Taking a page from the success of the *Sears Catalog* indirectly, major pharmaceutical companies needed to establish what only time and service would do. They also required a cultural currency to attract new business in a world where drug-related commodities were in every store, on every street corner, and inside every pharmacy and drug store. In order to attract attention European companies that now had outlets in America, like Bayer, Merck, and eventually American equivalents like Pfizer, Eli Lilly and Company, H. K. Mulford, Parke-Davis, and E. R. Squibb & Sons would capitalize on the ever-growing healthcare outlets that had continued to gather credence during the second half of the century. Bache's mid-nineteenth century vision of a credible scientific community in North America was slowly evolving, and a generation later those pharmaceutical firms recognized that a vast number of healthcare professionals would read and acknowledge this print culture as gospel.

In order to expand their chemical empire of opiates, firms recognized the expediency of advertising campaigns. However, it was complicated because as a company with a cadre of chemists and other academics in tow, it was still perceived as unseemly

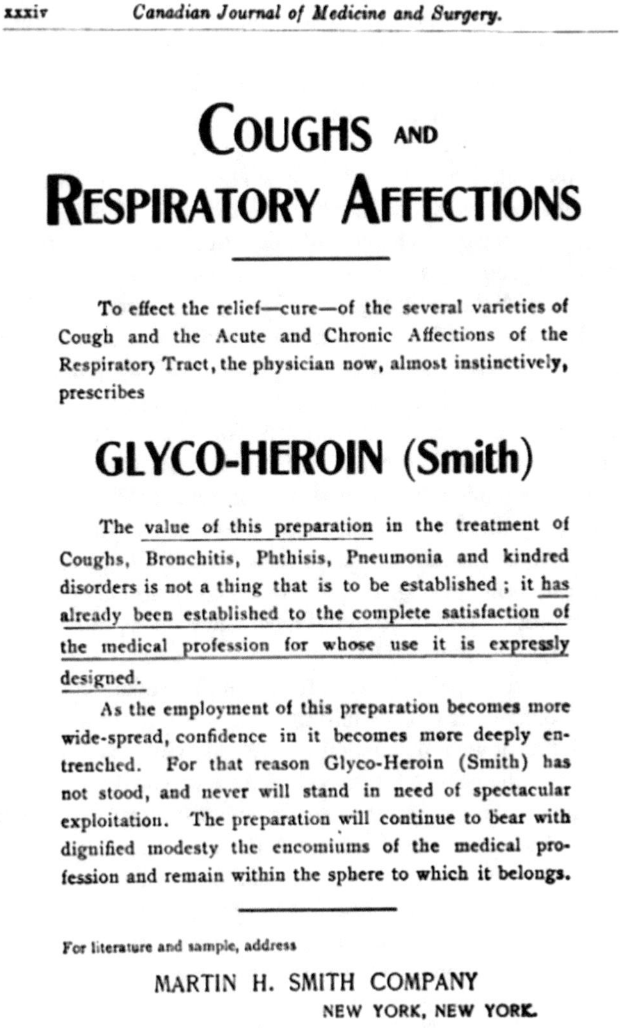

xxxiv *Canadian Journal of Medicine and Surgery.*

COUGHS AND RESPIRATORY AFFECTIONS

To effect the relief—cure—of the several varieties of Cough and the Acute and Chronic Affections of the Respiratory Tract, the physician now, almost instinctively, prescribes

GLYCO-HEROIN (Smith)

The value of this preparation in the treatment of Coughs, Bronchitis, Phthisis, Pneumonia and kindred disorders is not a thing that is to be established ; it has already been established to the complete satisfaction of the medical profession for whose use it is expressly designed.

As the employment of this preparation becomes more wide-spread, confidence in it becomes more deeply entrenched. For that reason Glyco-Heroin (Smith) has not stood, and never will stand in need of spectacular exploitation. The preparation will continue to bear with dignified modesty the encomiums of the medical profession and remain within the sphere to which it belongs.

For literature and sample, address

MARTIN H. SMITH COMPANY

NEW YORK, NEW YORK.

Fig. 3.4 Advertisement in *Canadian Journal of Medicine and Surgery* c. 1903. Photo courtesy of Early Canadiana Online

to engage in unscrupulous marketing. This was not the 1950s, nor was it the 1920s, at least not yet. In essence, how could a pharmaceutical giant wanting to not be seen as a patent medicine company working out of a back kitchen and offering advice through the post or viewed as a distributor slapping labels willy-nilly on product, still promote their business that they had worked so diligently to create? The answer was found in an extension of the laboratory. A place where the chemistry and the new biomedicine could meet in professional consultation, which would serve in an age before clinical trials (there was no drug discovery cycle just yet, per se), as the best means to disseminate information that could be helpful to patients. This forum was another incarnation of a chemical heterotopia, the medical journal.

Whether circulated in the form of dailies, weeklies, monthlies or on an annual basis, journals related to medicine have a long history, especially in Europe and were nothing new in the late nineteenth century. As previously discussed, apothecaries and pharmacies used them to express the latest knowledge, and local journals were at times self-published, and quite snobbish, since universities had yet to sponsor regular reviews. That began to change with the advent of university-based organic chemistry, but pharmacological publications continued on their own and in many forms. In America, with the founding in 1847 of the American Medical Association (AMA) and with their *Journal*, named after the organization, they attempted to serve as a clearinghouse for healthcare delivery methods, but they fell woefully short in translating the work of chemistry laboratories [14]. Recall, that the definition of a laboratory, once broadened, included nostrum cooks, whose products were extremely popular, albeit lacking in labeled contents. Drug disseminators, whether they were making alkaloids or not, plied their wares by any means necessary, so the market was like the quintessential nineteenth century food, the oyster, a world open to those that could crack it. Thus, the medical profession faced a difficult dilemma since all new medicines were tested only through practice, on real patients under actual conditions. Therapeutic revelations could be found in local newspapers, in letters to the editor or in a juried professional journal [15].[7]

The central question moving forward for scientific leadership, particularly pertinent to the development of opiates into opioids, was how could you promote the bountiful possibilities of what the laboratory created, while keeping medicine away from the corrupting influences of the market? Before we can discuss what they did to solve this issue, first it is important to examine how people received medical knowledge by 1900 and beyond. In most instances, readers of any form of periodical or journal learned about what a drug like heroin could do, the recommended dosages, whether it had side effects, and how different age groups reacted to its use. Normally, the author would encourage other physicians to try it on their own patients. So, in essence, an old game of telephone ensued in the hope that the message or in this case, the diagnosis, would take. In the event that the drug was a failure, messages would be sent out in several forms; likewise, if it was a success (as we saw with Pinkhams), endorsements rang true. Despite new forms of chemistry dominating the trade, individual patients and doctors still exercised control, since it was the latter's obligation to apprise themselves of the latest pharmacological developments [16]. Pharmacies were still very much part of consultations at this point, even more so in some places than doctors, and it was a set criterion that demanded an item in the public domain have a Latin attribution or public name, competitively available from a professional drug company, and additionally were never advertised to laymen [17]. Of course, as the AMA soon learned, this was not going to be easy to discern.

---

[7]Just a short piece of oral history, I recall my Mother and Grandmother cutting out newspaper articles and sending them to one another about the latest innovations in drug creation. *Ask Dr. Gott* was a favorite of both.

Fig. 3.5 The Merck Report Masthead. Photo courtesy of the Othmer Library, Chemical Heritage Foundation

As for those corrupting influences that came from the market, different laboratories addressed this issue depending on their intentions and the mission of the firms. Larger pharmaceutical firms that developed alkaloids, and specific cough suppressants like Dreser's heroin or Merck's European or American bases, had their own approach to promoting themselves. Merck in particular even went so far as to found a compendium of pharmacological and chemical knowledge through the publishing of a regular offering of what became known as, *The Merck Manual*, *The Merck Index* or *Merck's Index* as it was known, and *The Market* (Fig. 3.5).[8] Each well-organized text informed the professional chemist, the pharmacist, the doctor, and the lay person with the latest information concerning chemicals, drugs, and biologicals. With such succinct reading, that was much more accessible than the *United States Pharmacopoeia* (*USP*), Merck publications were not as technical for the average member of the public. *The Market*, in particular, was structured for wholesalers who needed to know the latest import rates and amounts. Opium was regularly listed, and interestingly enough by 1908 reflected a downturn in importation amounts [18].

With such useful knowledge now in plentiful amounts, the pharmaceutical firms moved towards a two-fold approach for their alkaloid platforms, and it led to indirect investments in medical journals. The first aspect of their approach was sending those all-important samples directly to medical elites who in turn would plug into their base and give it a run by publishing their unscientific results in regional or national publications. The second approach was backing up these recommendations with an actual advertisement that would be strategically placed in the same issue [19]. This two-fold approach placed authority squarely in the laps of the attending physician, so to speak, and reflected the art of medicine and deft advertising. If well-received doctors acknowledged a product, then it would gain acceptance. It was the best-case scenario for a company attempting to promote its drug manufacturing capabilities,

---

[8]*The Merck's Index* was in print from 1889 to 2012 when it was acquired by the Royal Society of Chemistry. An online version is available through research libraries.

while walking that fine line between avarice and rectitude. An example from 1903 pertaining to heroin is instructive.

Dr. John Leffingwell Hatch was a leading physician of the early twentieth century. Seemingly everywhere, he had all sorts of connections in both America and in Europe. Born in 1863 in Rochester, New York during the American Civil War, he trained at the University and then did a post-doc at the University of Pennsylvania. He probably heard about the newly finished Chandler Chemistry Building in Bethlehem in 1888, the same year of his graduation. Like those exo-chemists that trekked to Europe for training, Hatch followed suit, and landed in Heidelberg, Germany, where he studied at university and in several biologic laboratories. He had some German in his repertoire from his days at Rochester, plus he developed an interest in crafting poetry and short stories, which the former probably helped with the language barrier more than the latter. Traveling just south of Merck's Headquarters in Darmstadt, he returned to America obtaining numerous fellowships and serving as a surgeon for steamship line and the United States Marine Hospital service. In short, he amassed a strong resume.[9] By 1902, he was well-travelled and had developed a dossier of experience that led him to regularly publish his treatment methods in numerous juried medical journals, which made him rather a celebrity when it came to medical appraisals. With a connection to German pharmaceutical firms he was well-placed to serve their needs. And, as far as we know, they did not even need to compensate him [20]. In several journals, including the *Canadian Journal of Medicine and Surgery*, a publication that was distributed widely across that country and in America, he published in the October 1903 issue the article hauntingly titled, "Glyco-Heroin (Smith) Almost Infallible." Based on evidence he had collected from some "fifty cases" concerning patients he had encountered that took Glyco-Heroin, he reported fabulous results. He found patients receptive, of all ages, and in his professional opinion, there was much hope for it. What is most instructive about this source is not necessarily that Hatch thought Glyco-Heroin was an excellent product (because we now know that it was not, especially for children); rather, what was most significant was what other journals did with his article once they saw it published in the Canadian journal. Like wildfire, Hatch's recommendation for heroin as a cough suppressant spread quickly. The article or a reference to it could be found in regional journals like the *St. Louis Medical and Surgical Journal* to national ones like the *Journal of the American Medical Association* (just to name a few), which conceived of itself as the premier

---

[9]No evidence for or against remains on whether John Leffingwell Hatch (1863–unknown) was ever on the payroll of Merck, Bayer, or Glyco-Smith. He appears to be quoted in numerous publications, and since copyright infringement was still being developed as a branch of law it was difficult to prosecute cases or discern whether these firms received the approval to quote Hatch. It is not out of the realm of possibility that Hatch developed relationships with Merck and Smith since he was near both Darmstadt and New York during his career.

publication for all things medicinal.[10] Within a month, Hatch's appraisal of Glyco-Heroin leapfrogged by receiving broad coverage and probably made Smith, and for that matter Bayer, extremely pleased that such an eminent surgeon (notice Hatch's degrees are listed after his name, which speaks to his experience) as he would publish such an excellent testimonial [21].

Despite Dr. Hatch's ringing endorsement, there was more that was required to complete the circle. Firms believed that visual print culture was just as persuasive as the written. Thus, Bayer, Merck, and even Smith rolled out new commercial announcements (Figs. 3.2, 3.3, 3.4). Taking out a full-page advertisement, which was not inexpensive, each vied for a one-two punch that would entice and create a secondary wave of sentiment concerning how to effectively suppress a cough. Even though Merck did not have their own heroin-based product, they still tout their own types of alkaloids (a drug called Dionin) that could treat the same symptoms. By taking advantage of the scope and volume of the journal and the transference of knowledge from one to another, large firms found an ethical lever for increasing awareness and sales. From an efficacy standpoint it was a brilliant move for a new corporate age of chemistry [22]. The laboratory was once again extended into the pages of a variety of journals. Copies would end up in the mailboxes and in the libraries of America. This occurred just in time for the expansion of learning in public education, in universities, and in the building of some 2500 Carnegie Libraries, which were constructed for those previously socially barred from the luxuries of learning.

The new medical journals for a new century represented a major shift in expertise and translation of what the professional chemistry laboratory was producing. Private interest had officially merged with the scientific process, but as the example of Hatch's endorsement of heroin would prove, there was a tincture that commercial exploitation was unduly influencing the *educational machinery*, as the therapeutic reformer Francis Stewart called it in 1911 [23]. Despite this critique, experimental knowledge was now in reach of both physicians and pharmacists in ways that were not possible before the influence of the laboratory heterotopias brought them into focus. The laboratory in this sense made alkaloids more reputable, especially considering morphine had so many destructive side effects for those in pain. The AMA had tried for decades to codify standards for drug-making and failed due to the power of the nostrum market, despite banning in 1900 all advertisements that were deemed illegitimate. Now, with industrial labs proving they could deliver products to the market that were endorsed, for better or for worse, by elite professional doctors, the future of opiates would be secure for companies like Bayer. Or, so they thought. Events in places as far away as the Philippines related to cultivating of opium and the making of opiates would dictate another change in trajectory and alter the course of the road to opioids forever.

---

[10]Other regional journals that carried Hatch's appraisal of heroin include: *The Journal of Medicine and Science* (Maine), *The New Charlotte Medical Journal* (North Carolina), *The Trained Nurse and Hospital Review* (New York), *The Medical Bulletin* (Philadelphia), and many, many, others.

3.1.4 The Costs of Slaying the Dragon: An Orwellian Tour of Therapeutic Reform

The Rosengarten & Sons, Manufacturing Chemists (founded in 1822) never knew that within the friendly confines of their company they were producing a bevy of highly illegal substances in what the United States government after 1914 deemed as, narcotics.[11] After all before that, no one, except a bunch of rabble-rousing teetotalers, conservative doctors, and goos-goos (those that believed in a just form of urban leadership) in city governments, had really made the connection between alkaloids and criminality. Yet, right before the First World War, the connection between the two was linked. Back in 1855, the company was focused on ramping up production with the construction of a new chemical heterotopia at the corners of 18th, Fitzwater, 17th, and Catherine Streets in the 30th Ward of Philadelphia [25]. Although it paled in comparison to what Bayer and Merck had created in Germany and would create in New York, by American standards the steam-powered complex (Fig. 3.6) was the largest in the country and would be able to keep up with the demands for morphine and other pharmaceuticals that were required by the North as it fought to preserve the Union. Rosengarten entered the alkaloid business steadily and by 1876 they were so large and powerful that they even constructed a massive display in the United States Chemical Wing at the 1876 Centennial Exposition in their own backyard of Philadelphia that displayed their latest wares [26]. Their chemical production included a staff of some 80 hands that assisted in making quinine, morphine, and other chemical preparations that were cooked onsite in the L-shaped building. As you approached the block you would have smelled the results of their processes and the signage that read, "Rosengarten & Sons. Established 1822" near the entrance drew the eye. The building contained the office, packing, and manufacturing rooms for the firm, which was one of the oldest chemical manufacturers in America. It would not last though. Falling on hard times due to competition from European firms and new Progressive initiatives during the Era of Theodore Roosevelt, the manufacturing heterotopia went from being a leader in plant alkaloid and bromine production, to part of a chemical merger that spoke to the fact that the company had entered a

---

[11] At this point in our story of opioids you might have noticed that we have not used the word narcotic(s). It is not a word from the twentieth century rather it comes from the fourteenth. Webster's offers the following:

Definition of narcotic:

1a: a drug (such as opium or morphine) that in moderate doses dulls the senses, relieves pain, and induces profound sleep but in excessive doses causes stupor, coma, or convulsions

b: a drug (such as marijuana or LSD) subject to restriction similar to that of addictive narcotics whether physiologically (see physiological) addictive and narcotic or not

2: something that soothes, relieves, or lulls

Etymology of narcotic:

late 14c., from Old French narcotique (early 14c.), noun use of adjective, and directly from Medieval Latin narcoticum, from Greek narkotikon, neuter of narkotikos "making stiff or numb," from narkotos, verbal adjective of narcoun "to benumb, make unconscious," from narke "numbness, deadness, stupor, cramp" (also "the electric ray"), perhaps from PIE root *(s)nerq- "to turn, twist" [24].

Fig. 3.6 Rosengarten & Sons, Manufacturing Chemists, Philadelphia, c. 1876. Photo courtesy of the Free Library of Philadelphia, Print and Picture Collection

new phase in development when it was bought by Powers & Weightman in 1905. Movements such as these portended much for the chemical worlds of America [27].

The inability of Rosengarten's sons to continue their family-based chemical heritage suggests the changing world of pharmaceutical production at the turn of the century. A nexus was emerging in which the voices of protest over the nostrum crisis would gradually pivot to a new set of attitudes, which by 1906 would greatly influence governmental policies. Public intellectuals and scientists alike were on the same page when it came to their quest for a commitment to responsible and effective drug-making. By the 1870s they were poised to develop a new program called *rational therapeutics*, and it would have a major impact on the continued creation of opiates, and eventually opioids to the present day. Back in the late nineteenth century, this new approach originated not with organic chemistry departments, but rather in medical schools that were evolving with the times as well. A renowned physiologist with an international reputation, Claude Bernard greatly influenced professors at the University of Pennsylvania (Dr. Hatch's alma mater), and at Jefferson Medical College, two of the foremost large and small venues that turned out the next generation

Fig. 3.7 Claude Bernard,
French physiologist, c. 1860.
Photo courtesy of Edgar
Fahs Smith Collection,
Kislak Center, University of
Pennsylvania

of physicians.[12] Their brand of what they called, *physiological therapeutics,* theorized that the chemistry laboratory and its ability to study drug action was where a revolution in treatment could and should take place. A combination of German-led laboratory medicine when coupled with these new standards and practice would in their minds, push a revolution through reform [28]. In a sense, they were taking the test tube, which Liebig and his cadre of students had passed to those same industrialized laboratories, which they summarily abandoned in favor of capitalism. By 1900, a fresh group of faces, no longer donning top hats and little other forms of safety equipment, emerged from their carols spouting new forms of rational therapeutics. Here is how it went (Fig. 3.7).

First, with an intellectual program such as this you needed to look at clinical practice.[13] For centuries, the way doctors had addressed the doling out of medicine focused on the art of sense. That is, does the patient look and feel better after it is applied? If they do, then excellent, the clinical practice worked, and it is on to the next one. In this system, once you had accrued a certain number of cases (as we saw this in action with the report from Dr. Hatch's appraisal of heroin in the *Canadian Journal*), then your background and expertise expanded. In turn, at this point you could make deductions concerning the diagnosis of symptoms, the means by which medicine was administered, and the penultimate cure became incarnate, presto. This was deemed as an empirical remedy, which would be like starting with

---

[12]Claude Bernard (1813–1878) was a French physiologist and was one of the first to suggest the use of blind experiments to ensure the objectivity of scientific observations. He originated the term milieu interior, and the associated concept of homeostasis.

[13]Over the past two decades sociologists interested in STS scholarship have taken criticism for solely focusing on the inner workings of the laboratory; in other words, not taking into account the role played by clinical research. It is not the intention of this study to ignore clinical work; however, this study is focused on why the laboratory over time was steadily ignored in relation to the history of opioids.

a blank canvas, paints, and brushes and then improvising along the way as strokes are applied. According to the rationalists, it seemed scientific, but it was not, since what medical delivery was essentially doing was counting up the brush strokes and standing back from the canvas with a nodding gesture and an introspective noise that followed. Instead, what they began to spotlight were not the therapeutic agents that were applied, but rather the mechanism of action. The way to study this was scientifically in the laboratory using demonstrations based in cause, not symptoms from the clinic.[14] In order to undergo a revolution in medicine, clinical pursuits would need to be based in known pharmacological science; rather than on the feelings of the individual or the independent rights of the doctor. Thus, opiates and nostrums were not exceptions, and became targets for progressives who were looking for sources of society's ills [29].

The therapeutic reformers pushed for two lines of attack, which would become the guiding principles behind organizations like the AMA and eventually the government's Bureau of Chemistry, which was founded after the Pure Food and Drug Act of 1906. First, as a society, control of the introduction and promotion of drugs must be retained by a clearinghouse that would act in the best interests of patients and doctors. And second, a groundswell of support that was scientifically-based needed to be harnessed which could hold the medical profession to a set of very high standards. Thus, rationalists saw restricting drugs would keep confusion to a minimum and create a just form of therapeutic practice. Likewise, an enlightened profession would be capable, as a reformed profession, of reinforcing the mission of only using products that were carefully screened by the laboratory. At its heart, reform would invent the scientific drug screening and lead to the professionalization of physicians. Now, it was time to turn theory into practice; the outcome was uncertain and the stakes were high [30].

The gauntlet of practice was first assumed through decades of fits and starts by the AMA. An inward-looking oligarchy if ever there was one, the root issue for them centered on how to agree which official positions the organization would adopt. After 1900, they began to move in the direction of addressing specialists in the field by forming a security council of representatives that could inform and consult on the latest drugs on the market. Finally, in 1905 the AMA's Board of Trustees approved the creation of the Council on Pharmacy and Chemistry. Made up of individuals selected from those that had intimate knowledge of pharmacology, this new organization within an organization, would attempt to offer opinions and in the end, regulate drugs and their promotions. Probably the most influential member of the AMA that spearheaded the Council was a physician and medical journalist named George Simmons, who had strong opinions about opiates, the role of the laboratory, and how the federal government could take part in evaluating the latest medical commodities.[15] As Simmons viewed the situation his argument was that the new Council could utilize

---

[14]There were addendums to this approach. For instance, if there was no specific cure, symptomatic treatments would be allowed, but research still had to prove its process in achieving relief.

[15]George Simmons' (1852–1937) life spanned an amazing period of growth in medical history. Based in Lincoln, Nebraska, he came to the AMA after serving as secretary of the State Medical Society and as the editor of the Western Medical Review. He joined the AMA's Board of Trustees and crafted the mission moving forward for the organization by advancing the agenda through his

the *Journal*, just as it had by Bayer, Merck, and Smith to promote their cough suppressants, as sort of the last stop along the train track of drug creation. As a chemical heterotopia on paper, the JAMA would have a special section named, *The Propaganda of Reform*, which would seek to editorialize salient points of view and advise the profession on how to avoid quackery. By 1907, the JAMA converted this accrued knowledge into an annual installment of their own version of *Merck's Index*, with the publication of the first edition of the *New and Nonofficial Remedies* [23]. Taking aim at the cooks and their nostrums, Simmons' Council set as a goal to bring into the light those phantom ingredients that were so well-hidden from the public's purview.

Simultaneously, as the AMA pursued its mission to institute rational therapeutics, they inhabited one of the most concentrated periods of social action in American History. The Progressive Era inaugurated the end of the Gilded Age with an inspired, albeit restrained by modern standards, attempt at governmental intervention. The movement targeted political machines, where honest graft was the letter of the day, and their spirit also included taking on those trusts that Carl Duisberg had so admired when he visited the United States for the first time. National political leaders addressed change through legislative means. They supported issues like prohibition, direct election of Senators, women's suffrage, and scientific management through increases in efficiency. Engineering change was not easy politically speaking because the Republican Party, which had maintained a majority for most of the period since the Civil War, advocated industry as a means to wield great power. Controlling the pharmaceutical companies was not necessarily in their best interest, but the pressure exerted by the AMA, debates over an issue such as patenting, and the rise of muckraking journalism forced their hand. That final set of influences was particularly reflective of the period. Intrepid reporters looked at the slaughterhouses, the factories full of children, and the greedy tentacles of Standard Oil as a series of blights that were plaguing humanity [27]. Dubbed muckrakers, because they got down into the mud and the muck of corruptive practices, magazines and newspapers, like *McClure's* and *Collier's*, began to specialize in exposing the waste and scandals at the state and local levels. For subjects like health and safety, the middle class proved both a motivating and ready audience, especially concerning the state of nostrums. One journalist in particular proved that print culture could move the legislative football one hash mark at a time. Samuel Hopkins Adams, writing for *Collier's*, did the most damage when he published a series of pieces beginning in 1905 entitled, *The Great American Fraud*. Selling over 150,000 copies over several years, the critique covered the evils of patent medicines and was particularly scathing. Shocking revelations were related in the *Ladies Home Journal* and a host of other periodicals that banned advertisements of nostrums [1].

From 1906 until 1914, several changes arrived via the federal government that would alter the overall trajectory of the nostrum crisis and the production, sale, and consumption of opiates. Reformers and the scientific laboratory received a major boost with the passage of the Pure, Food and Drug Act of 1906. After two decades in

---

editorship of the *Journal of the American Medical Association*. Serving for 25 years, he saw the subscription rates increase from 1000 to 80,000 members.

a quest for parity in food and in drugs, and especially in an attempt to curb opiate use by defenseless children, the revolutionary act was the first of its kind. The newly created Bureau of Chemistry (BOC) five years before was directed by the Department of Agriculture and oversaw administratively a vast set of issues in the manufacture of pharmaceuticals. Under the direction of the German-influenced and exceedingly capable Harvey Wiley, who was instrumental in providing the research for the bill itself, and his handpicked director, Lyman Kebler, the BOC as they conceived it, would serve as a multifaceted and valuable guardian against the abuses of the unregulated and undisclosed patent medicine trade.[16] Founding four different labs before 1910 that comprised drug inspections, synthetic products, essential oils, and general pharmacology, Kebler struggled to bring together a large consortium of knowledge under one roof.[17] It was ambitious, but wrought with challenges, as companies found loopholes in the Pure Food and Drug Act's language [31]. As long as they printed names of the contents on the labels, they could slip past the BOC. Companies founded complicated names that the average consumer could barely pronounce, which created confusion and frustration. Still, Wiley's laboratory heterotopia would be the first government-backed entity of its time, and draw from the latest pupils of the organic chemistry, industry, and rational therapeutic generation. Invoking Taylorism, which called for a progressive spirit of engineering and design, the future looked bright for the laboratory as a place that could help to realize Liebig's Dream of social consciousness and chemistry-for-all (Fig. 3.8).[18]

The Pure Food and Drug Act created a grocery list of specific items including alcohol, morphine, opium, and heroin that were required by law to appear on labels of

---

[16]Harvey Wiley (1844–1930) was born on a farm in Indiana until he enlisted in the Army serving as a corporal for the North. After studying chemistry at Hanover College and receiving an M.D. the Indiana Medical College in 1878, Wiley, like many others, travelled overseas where he attended the lectures of Liebig's greatest student, August Wilhelm von Hofmann. While in Germany, Wiley was elected to the prestigious German Chemical Society founded by Hofmann. He spent most of his time in the Imperial Food Laboratory in Bismarck working with Eugene Sell, mastering the use of the polariscope and studying sugar chemistry. After a long career in government his political enemies forced him out, and he went to work for *Good Housekeeping Magazine* in their consumer department and laboratory. His main cause—the dangers of consuming too much caffeine.

[17]Lyman F. Kebler (1863–1955) graduated from the University of Michigan in both pharmacy and chemistry. He then worked in industrial drug development in Philadelphia for the firm Smith, Kline, and Smith as their chief chemist in charge of testing potential materia medica for investment. At the BOC he spent the first two years testing opium assays for morphine. This is interesting because it runs along two lines of inquiry: one, the medical and pharmacological fields who were grappling with opiates, and two, the federal government, which was pursuing political and economic interests with opium in East and Southeast Asia.

[18]Taylorism, named after the US industrial engineer Frederick Winslow Taylor (1856–1915) who in the *Principles of Scientific Management* (1911) laid down the fundamental principles of large-scale manufacturing through assembly-line factories. He emphasized gaining maximum efficiency from both machine and worker, and maximization of profit for the benefit of both workers and management. Although rightly criticized for alienating workers by (indirectly but substantially) treating them as mindless, emotionless, and easily replicable factors of production, Taylorism was a critical factor in the unprecedented scale of American factory output that led to Allied victory in Second World War.

Fig. 3.8 Harvey Wiley,
early in his tenure at the
Department of Agriculture c.
1893. Photo courtesy of
Edgar Fahs Smith
Collection, Kislak Center,
University of Pennsylvania

any pharmaceutical, whether it was a nostrum or otherwise. Charges of misbranding, which carried prison sentences, could be leveled against a mom out back in the kitchen or a major chemical manufacturer. Almost overnight, the kitchens closed or relocated to other more secretive heterotopias. The first illegal labs producing a host of concoctions were born. Organizations like the AMA felt as though they had hit the jackpot, but the era of regulation and oversight was only the beginning. For Wiley in typical fashion, actually thought the legislation had not gone far enough. Still, it gave the BOC the authority to define the relationship between *names* and *things*, as the wording went, which determined in the end what was and was not *adulterated*. In essence, anything that bore *false* or *misleading* statements was subject to being pulled from circulation. With federally-sponsored publications at their disposal, and coupled with the editorials found in the JAMA, coverage of a muckraking quality could be leveled at any suspected perpetrators.

As mentioned, the era was not complete since right before Europe was plunged into what would become the first truly massively destructive global war, the United States created the first modern narcotic, simply by naming it as such. The Harrison Act, as it became known, was the first federally-supported law which regulated and taxed the production, importation, and distribution of opiates and coca products [32].[19] It was sponsored in the House by New Yorker Francis Burton Harrison, an avowed Filipino-file who had seen firsthand during the War of 1898 how utterly ruthless the opium trade was and what it did to people who smoked it. Prior to 1913, the Roosevelt Administration sought to control the Philippines as an imperial prize

---

[19]Cocaine is not considered a narcotic nor an opiate, but it was placed on the Harrison Act's watch list since particularly in the South it was seen as being abused by all kinds of people, especially African Americans (who were segregated and always suspected stereotypically of criminal activity), but also by those of many different socio-economic backgrounds.

that was justly won after Dewey's glorious capture of Manila and the defeat of the malevolent Spanish. Standing in for them, the Americans incited a revolt when Emilio Aguinaldo and his freedom fighters were beaten and tortured (made to drink copious amounts of water, then had their stomachs stomped on) by the very conquerors they had assisted. In an act of benevolence, but also self-interest, the President convened an International Opium Commission in Shanghai, China in 1909. The result was the appointment of one Hamilton Wright, a doctor from Ohio who had trained at McGill University in Montreal and had done a few stints in China and Japan. Wright's mission, as an opium czar, was to pair with Roosevelt's handpicked successor to the White House, the rotund William Howard Taft, in an attempt to elbow America's way into the feeding frenzy that surrounded the impending fall of the Qing Dynasty in China [33]. They found a gateway by making opium look like a primary concern because of the number of addicts that could be found in America. By crusading against drugs first in the Philippines and then at home, they could force China to come to the treaty table and negotiate a new set of trade contracts. Drugs, certainly not the first, nor would they be the last, were used as a lever for international politics and trade concerns. In order to tackle the runaway opiate crisis, enter the classic reaction by a bloated government and bureaucracy seeking to do good in one fell swoop, the Harrison Act. By declaring importation illegal, it served the dual purpose in the United States of trying to end the exchanging of the sticky sap (thus, opening the door again into a China by 1911 that would become a proto-Republic under strongman Yuan Shikai), and also castigate addicts who were thought of as criminals, as a check against the progressive spirit of the age [32].

Just before 1914, the United States looked like a nation that had slain a dragon. Nostrums were becoming less fashionable, narcotics were named and banned, and the AMA was ascendant and professionalizing. All looked promising with the federal government's new chemistry laboratory (BOC) employing the best and the brightest under the command of a man who knew policy and practice. The most aggressive gain was how rational therapeutics had pushed medicine to reject the impulse to believe in covetous market-driven pharmaceuticals that were untested. All of these points seemed overwhelmingly positive for fields best described as an absolute mash of perspectives, products, and pathologies that woefully served the patient. Individuality, which was the hallmark of doctor's prescriptions and advice, and part of a patient's right to self-medicate as they saw fit, seemed at an end. Yet, despite these victories against the abuses and misuses of opiates, a seed was planted that went unnoticed by the powers that were. Instead, nostrums went underground to fight another day. Narcotics in the future would come in legalized doses from the hands of the pharmacists, and written by the pens of doctors still artfully seeking to alleviate their patients' pains. The Harrison Act in actuality created more paperwork for those physicians who saw this as an inconvenience and as a bureaucratic nuisance for their practices, rather than a method of therapeutic codification. Finally, last but not least, steadily the professionalized chemistry laboratory, which showed such promise as the origin and purveyor of rational approaches to medicine, would become more central to developing new drugs, but in the same right more susceptible to the trappings of commercialization; as Big Pharma heterotopias built complex means to influence the very legislative gatekeepers that sought to protect an innocent and unsuspecting public. It was an Orwellian tale in-the-making.[20] Big Pharma in the United States was initiating a different type of chemical network, taking advantage of graduates from different disciplines in a new age of biomedicine, that would make research and product development the province of the board room and not the laboratory. This arrangement created a virtually impregnable mode of development for an industrialized planet that was sprinting faster and faster in order to keep up.

3.2 Roots of Modern American Big Pharma Takes Hold

First, we need a fully equipped chemical laboratory, then a pharmacological institute with a staff of men trained in medicine and chemistry, an abundance of animals to experiment upon… equipped according to the ideas of Paul Ehrlich; all these must be in close connection with one another.

—Carl Duisberg, c. May 1913 [34]

My chemists and I deeply regret the fatal results, but there was no error in the manufacture of the product. We have been supplying a legitimate professional demand and not once could have foreseen the unlooked-for results. I do not feel that there was any responsibility on our part.

—Dr. Samuel E. Massengill, Owner, Massengill Company, Bristol, Tennessee c. 1937 [19]

3.2.1 The State of the Chemical Teutonic Nexus, c. 1914

Right before the August guns roared, sounding the death knell for humanity itself, Bayer pulled heroin from its production line. There are several ways to interpret this decision. Perhaps Carl Duisberg and the Bayer leadership believed they had a fiduciary responsibility to their customers and the doctors who prescribed it. Another possibility we could surmise was that the Board of Directors felt that it would be a poor reflection of their business strategy or maybe that if so many were turning into junkies this would turn public opinion against them. Some scholars have purported that since aspirin was doing so well and becoming a global wonder drug, that Bayer did not need controversy since profits were so high. Whatever the case Bayer's action had both short and long-term consequences. In the short, it and the subsequent passage of the Harrison Act, pushed heroin into the underground realm of criminality, and in the long term proved that a pharmaceutical company could exercise good judgement when it came to how their products impacted society. The caveat was that they had to listen, which was an extremely difficult admission in the face of maximizing profits. The period of 1914 to the end of the Second World War in 1945 was not just a period of history where Germans collided with Britain, France, Russia, and America in two conflicts. It was a referendum on the history of the modern world and the future course

---

[20]Orwellian is defined as, of or relating to, or evocative of the works of George Orwell, especially the satirical novel *1984* (1949), which depicts a futuristic totalitarian state.

it would take. Death, destruction, disease, the very definitions of race and empire, and even science were all part of the larger story in the race for hegemonic power. From a chemical perspective, industrialism and design fueled new means by which to enact violence, but there was this paradox that ran concurrently. Biomedicine struggled to wage its own war against pain and pathogens, all in the face of death from bullets and events like the Great Pandemic Flu of 1919 [35].

The social and intellectual landscape of medicine and chemistry during this period included a continued struggle not just between the two, but across and in between platforms. Medicine disagreed with biochemistry departments, and industrial firms quarreled with medical associations in different laboratory heterotopias. Ironically, like the German doctor Ferdinand Maack's Raumschach game of 3D chess which became popular after 1907, these different fields engaged in a struggle for souls as well as marks, pounds, francs, rubles, pesos, dollars, and any other forms of currency. The difference between consumers and patients continued to embrace a blurred chasm, especially when it concerned synthetic opiates and pain relief.

In the wake of Bayer's choice to discontinue sponsorship of heroin, a revolution in medicine and chemistry was being forged. Medical schools, which in the century prior were little more than certificate programs and rubber stamps for physicians, were now producing those that might not practice research, but believed deeply in its benefits. This new generation of doctors and administrators in new spaces, like the professional modern hospital, would be the ones to appoint the next group of Ph.Ds who were schooled in physiology and biochemistry. In turn, they would do stints in clinical settings and then shift to industrial laboratories. Vacillating between each of these entities (the clinic, the firm, the university) allowed for cross-fertilization, albeit a hectic set of fits and starts [36]. Even though medical chemists did not specialize at this point in physiological chemistry, the opportunity would arise after 1945 in the form of the applied sciences. Liebig's Dream would have to wait until then.

In the meantime, an interesting nationalistic brotherhood developed between Germany and the United States which would form bonds that could only be loosened with an international crisis. Along two parallel tracts, these powerful modern industrialized upstarts represented the very best innovators when it came to the future of opioids. The decisions they made chemically and medically would reverberate through the development of drugs like pethidine, methadone, oxycodone, and fentanyl. A chemical lineage was forged during the rise of the German state in the 1870s and also in the Gilded Age in America, as professors, chemistry students, doctors, and the like, travelled and made those important connections with one another. The Americans in particular, since they became the victors in war, would pick and choose the portions of the chemistry laboratory that would suit their needs to address pain through a new biochemistry after 1945.

First, the chemical development of Germany had the unique history of being geographically blessed, able to draw on a pharmacological heritage, and once unified, the chemical industry enjoyed unprecedented power as a business since cartel-like companies were in the hands of the few. Before 1914, as in America, firms that did not band together were buried by the competition. This situation in actuality placed less of an emphasis on capitalistic pressures and allowed applied science to become part of the everyday lexicon of university chemistry and medical departmental programs. Thus, Germany possessed a dynamic pharmaceutical industry that was stratified, scaled, and integrated, instead of disparate, fragmented, and full of conflict [37]. The reason for this construction was Germany was the first nation to understand that in order to become leaders in the laboratory of applied sciences, it would be necessary to link academic medical scientists, practitioners, government officials, and producers of the new therapeutics around the same table. There just needed to be an entity that could tie them all together. One voice would be necessary in order to innovate. Luckily for Germans they had a person who possessed both a singular vision for those applied sciences and a penchant for how to mediate commercialism and research. They had Paul Ehrlich (Fig. 3.9). As a biochemist, Ehrlich would have a profound impact on not only biochemistry, but also a host of other issues including commercial science and research administration. His vision of a world of pharmaceuticals would reverberate throughout the twentieth century to the present day.[21] Ideas such as fixed dosage combination drugs and the laboratory as the progenitor of chemical knowledge in which all else would follow would be massively important to the future of turning materia medica, an outmoded phrase, into pharmaceuticals. In essence, time release drugs and nano-technology were all ideas he hinted or predicted would come to pass [34]Ehrlich's future placed an emphasis on theoretical medical science that would initiate a product-development laboratory oriented towards connections to the very best firms in commercial chemistry. Rechristened the Paul Ehrlich Institute after the Second World War, the quasi inter-disciplinary hub would direct and constantly reassess chemical practices and procedures.[22] If for instance, as was the case with heroin, a drug was being exploited, mismanaged or proved clinically dangerous, an overhaul was in order. Without the constraints of bureaucracy, centralization would cut through red tape and transmit the latest knowledge through the network; fiber optics, without the fiber. Large-scale testing functions would look at volume-control and examine how theories could effectively be turned into practice [34]. Ehrlich and his team at the Institute focused their attention beginning in the 1880s on alkaloids and their fever-reducing effects. This in turn evolved to examining a cure for tuberculosis, which Bayer honed through Dreser's process in the development of heroin. With his colleagues forming a chemical circle of ingenuity, they continued

---

[21] Paul Ehrlich (1854–1915) made a number of important contributions to medicine and chemistry in a career that multiple decades. He won the Noble Prize for Physiology and Medicine in 1908 for his combined work with the Russian immunologist, Ilya Mechnikov. Ehrlich's work in bacteriology and experimental pharmacology were based in establishing a framework for a modern research system within the laboratory. He is probably best known for the magic bullet concept for the synthesis of antibacterial substances and the development of chemotherapy. His understanding of drug-making would lead to theories associated with the development of antibiotics and the modern-day notion of the more drugs utilized in the proper combinations, the better.

[22] Initially called the Königliches Institut für experimentelle Therapie was modeled on the Pasteur Institute, but its links to theoretical science and commercial entities was much stronger and well-established, especially when it pertained to serum development. For our purposes in this study, referring to it as the Paul Ehrlich Institute denotes his influence, even though it was not named as such until after his death in 1915.

Fig. 3.9 Paul Ehrlich, c.
1910. Photo courtesy of
Edgar Fahs Smith
Collection, Kislak Center,
University of Pennsylvania

to press the field to think of themselves as distinct yet unified in purpose. With every
potential product immediately patented, unlike in America, German firms marked
their territory with certain drugs that treated specific maladies, which made each of
them unique. Pharmaceutical companies would collaborate and their testing facilities also served the governmental structure. When the central government in Bonn or
in Berlin needed data or to consult on issues, such as military hygiene, they turned
to Ehrlich and his team for answers because they knew he specialized in the latest
laboratory techniques. Private, self-governed, and structured for administration and
commercialism, the hub would find the answers in order to improve efficiency [34].

Paul Ehrlich's vision was not one based in traditional chemistry. For him, biological investigations linking the laboratory to questions like how can we create a
pain relieving synthetic that does not have the side effects of morphine or the habit-
forming qualities of heroin, was the proper approach to these types of questions.
This was a fundamental alteration in practice from previous investigations because
this type of chemical heterotopia would be fully integrated in order to make the
best resources available in the service of medicine. As a translator and intermedi-
ary between academia and industry, biochemistry would have the chance to flourish
under the umbrella of applied science. Physicians, previously held captive by chemistry, could take solace in the fact there was an enlightened set of checks and balances
in their corner, which would keep producers and advertisers at bay. Instead of drugs
being tested after they hit the market, now a proactive system could use biological
investigations to determine pharmaceutical solvency [37]. This was precisely why
Bayer pulled heroin in 1913. They knew that the Ehrlich Institute would pressure

them to remember their responsibility to best practices and an advocacy for those that were patients first and consumers second. This was the complete reverse of what was going on in America.

Recall that the industry there was dominated by heavy chemicals. It was not until the expansion in the production of other materials that developments in different types of alkalis were produced. A 1912 study by the Tariff Board was particularly telling. It appears that ninety-eight percent of applications for patents in the chemical category were assigned to German firms and amazingly, none of those were ever used in the United States. Though seven firms made synthetic dyes in 1914, it was actually the German intermediates that filled the gaps [38]. Meanwhile, the Americans simply did not have the technological savvy nor the experience to produce organic synthetic chemicals. Specialized pharmaceutical chemicals like alkaloids made by family-owned chemical companies were the extent of their development. Despite the founding of the BOC, by 1914 Harvey Wiley was at *Good Housekeeping Magazine*, and the efficient nature and spirit that existed during his tenure began to wane. The Harrison Act looked progressive on paper, but despite the fact that 35,000 doctors were indicted for prescribing narcotics, it did little but cause more paperwork for overwhelmed physicians. The 1906 Pure Food and Drug Act turned out to be a mirage since drugs could not be pre-screened and prosecution of those that refused to list ingredients could not take place since false claims had to be printed on the packaging [7]. As mentioned, there was tension in America between the different chemical heterotopias. Unlike Germany, the United States was obsessed with a clinical past that viewed laboratories as manipulated environments by greedy firms. Everything to these iconoclasts seemed like a nostrum. Within the clinical world, old school rivaled the new, as medical schools continued the growing pains of which forms of science they would learn to apply. Would it be the new biochemistry? Or forms of organic chemistry? How would this be carried out in clinical settings? Thus, their graduates would really need to master an applied approach to the sciences in order to serve their patients best. The reform movement in America became an important negotiator for both sides because one issue they could agree on was that undocumented patent medicines and opiates needed to be restricted [27]. Radicals, like the Christian Science Movement founded by Mary Baker Eddy in Massachusetts in the late nineteenth century, also helped when the public heard about children who died when their parents refused to use medical treatment for maladies like diphtheria. The famed reformer Abraham Flexner chided both sides for not coming together. He entreated the pathologist with his microscope and the clinician with their stethoscope, to look at the lab and the bedside as part and parcel of the same commitment [23].

If biochemistry could not gain a foothold within the university departments and in the industrial firms of America, one place it did flourish was the development of the clinical laboratory. A type of chemical heterotopia unseen before in Europe, this was a recent development in 1910 as the hospital laboratory, a new architecturally significant structure, joined the landscape of biomedicine. During the late nineteenth century, the clinical lab was a place mostly associated with the autopsy room under the auspices of pathology departments, which were connected with city and county governments [39]. By 1914, the pathology labs had evolved into important and complex institutions as doctors relied on their testing and for consultation concerning the dissemination of important pain medications. Clinical laboratories also became places where experimental and research methodologies were developed. If universities or firms refused to share research with the clinic, they would stage the lab under the same roof. This era did not last since America's entry into the First World War in 1917 drained laboratory chemists, and the hospital lab reverted to a testing facility. Still, it would serve as a portent for things to come after 1945 when the laboratory within the hospital would once again flourish in advocacy of biochemistry.

The root issue for clinicians in America was the question of resources. To them, a cultural and material image espoused in their minds the nineteenth century ethic of chemical studies led by the persevering individual scientist. American researchers looked at the example of Paul Ehrlich who screened and perfected laboratory science through the examination of thousands of alkaloids. Of course, this was not the reality for Ehrlich, who worked alongside a complex network of chemists and clinicians. He was never alone. It actually took America's entry into the Great War in order to bring cooperative chemistry into a practice such as this. The rising star of the Republican Party, Herbert Hoover, in charge of the Food Administration, believed an associative state could bring more scientific efficiency and produce political advancement as well [40].[23] Engineers across the scientific spectrum saw the Great War as an opportunity for advancement in material and organization; thus, inaugurating a new perspective for an age where killing fellow human beings ran congruently with exploring the means to save them. From 1917 to 1918, the National Research Council and the Carnegie Institution promoted chemical exchanges between their laboratories, and the federal government tried to keep up with demands for morphine and medical supplies to send with Pershing's army. However, the lack of infrastructure proved too much even for Hoover, and American doughboys had to borrow pretty much everything from the Allies once they landed in France. Cooperative science, like the advent of research-oriented clinical labs, would have to wait [41].

Why was America so deficient when it came to integrating the laboratory into one voice that could solve the problems of nostrums or something with fewer side effects than morphine? Some of the answer resides in the fact that America had placed an emphasis for too long on individuality. No matter how much they admired German pharmaceutical development they could not shed the idea of sacrificing intellectual autonomy. After the First World War though Germany reeled from the Peace at Versailles and its punitive reparations, through the efforts of Carl Duisberg and Carl Bosch, they combined their chemical strengths through the creation of IG Farben [42]. Back in the United States, due to trust-busting, those types of cooperative efforts and spirit were completely stymied. Though the land of the free for most, the arch of evolution in both medicine and chemistry each possessed inhibitions and prized above all else their own judgement. By 1918, the only means by which the United States

---

[23]To expound on this, the *associative state*, refers to Hoover's belief that responsible cooperation between governments (national, state and local), the American people (consumers/workers), and corporations would advance the economy and the standard of living in America. He believed in ability of all Americans—regardless of race, color or creed—to take care of themselves (through individualism) if given the opportunity to do so.

could possibly level the playing field against the German pharmaceutical juggernaut was to somehow have them undergo a major political, economic or military collapse. They found that opening in Paris during 1919.

3.2.2   An Alkaloid Empire, For Sale: The 25 Minutes that Saved Merck

Carl Duisberg realized a dream, possibly bigger than he could have ever imagined; one that only a vainglorious and low-functioning sociopath could squander. Though Germany decidedly lost the First World War, they still made major contributions to the future of pharmaceutical creation, manufacture, and testing under the watchful eye of the newly fashioned IG Farben during the 1920s and 30s. How they did this during such a massive undertaking of manpower as the Great War and the subsequent losses they sustained during and after is quite staggering. Also, recall that the destructive global Pandemic Flu of 1919 followed the death and destruction of the Great War. From a counterfactual perspective, it makes one wonder what they could have accomplished without the long-term decisions of the Kaiser and his generals who were involved in the destruction of their own empire? Or short-term ones like that blank check that was supposedly written for the Habsburgs, who incidentally were on their way out anyway. The end of the war to end all wars was incredibly debilitating to Germany as historians have studied ad nauseam, but at home the landscape was virtually untouched since the battles had taken place along the Hindenburg Line in France and Belgium [42].

During the struggle German companies that had forged both separate and close relationships with their parent firms were summarily cut off and auctioned to the highest bidder. Bayer's Rensselaer Plant and their entire aspirin empire was a case in point when it was handed over for a song to a bunch of West Virginia hucksters. Under the Alien Property Custodian (APC) all German assets now belonged to the American people, and would be dispersed accordingly.[24] After undergoing all sorts of anti-German sentiment, which ranged from renaming sauerkraut, *liberty cabbage* to suspecting the Kaiser's companies as part of great chemical plots, the seizure of chemical firms marked an important turning point in the history of technology, science, and for capitalistic economy geared to a future associative state [43]. In a sense, the United States was gaining not only the chemical know-how that came with these entities, but also the ability to develop future corporate structures that when the time was right would in turn produce new heterotopias that could not exist before.

---

[24]The harshest policy aimed at the German chemical industry originated from the Office of Alien Property. Created through an Act of Congress in October 1917, the APC, as it was known, derived its power from the Trading With the Enemy Act, which ordered the APC to manage the property was seized under the law. Persecution of Germans, whether first generation or not, was punitive, and there were all sorts of examples from public hangings to cultural changes like altering street names. This xenophobic spirit spilled into a controversy surrounding the making of phenol, which was associated with Bayer and other pharmaceutical companies.

Once President Woodrow Wilson returned exhausted from his sales pitch for the League of Nations, one of the few concessions he scored at Versailles, the APC readied itself to cash in on all of those German-owned patents that the Tariff Board had surveyed back in 1912. A. Mitchell Palmer headed the office beginning in 1917. He was a prickly villain who believed that the United States should not avenge the deaths of the 128 Americans who died when the Lusitania was sunk by a German U-boat in the Atlantic in May 1915 because it was not worth the lives of citizens. His progressive spirit led him to abuse his power at the APC before becoming even more famous as Wilson's new Attorney General in 1919. In that position he inaugurated the first Red Scare during his raids against what he deemed as the new threat of Communism. Through the dissemination of some 30,000 trusts (which included pharmaceutical companies) with assets worth the exorbitant sum of $500 million, he estimated there were an additional 9000 that awaited evaluation. For Palmer, war granted a nation certain inherent powers that transcended business connections. He was convinced that German industry sought to rule America, and everything from beer to chemicals, served as avenues by which they would accomplish this hostile takeover [44].

Interestingly enough, small businesses across the land saw through his clever scheme of auctioneering and cried foul that German monopolies would only be replaced by American ones (one American company secured 1200 patents in point of fact). Chemically-speaking, it was Palmer's subordinate and successor who concocted the idea of a non-profit with a suggestion that would raise revenue, defend patriotism, and increase innovation. Francis Garvan, equally suspicious of German chemical power as Palmer was, proposed that a private foundation be created that would hold the patents in trust and allow American companies to obtain licenses on a non-exclusive basis. The result was the Chemical Foundation, which over the course of ten years would oversee the sale of hundreds and hundreds of patent licenses; thus, raising the fortunes during the 1920s of some of the most powerful chemical companies in America, including Du Pont and Bakewell, just to name a few [45]. With a high protective tariff in place that would rebuke any coal-tar chemical competition from outside the United States, the newly powered science-based corporatism of an industry that seemed so far behind in 1917, was steadily gaining strength. Wars always provide lessons, and spawn new means of thinking and action. For the chemical and specifically, the pharmaceutical industries in America, they were reborn in what seemed like an overnight conversion from the days of family-owned and privately financed operations. Now laboratory sciences had the financial backing coupled with large-scale continuous operation. Research translated to the development of financial success. Finally, the Americans had arrived at a crossroads and a new chemical heterotopia, known later in the century as Big Pharma had officially arrived. The case of Merck is particularly instructive and its rise from the ashes of its German antecedents would have a major effect on the development of opioids in America.

While these events unfolded, George Merck was doing everything under the sun to become an assimilated American; culturally, politically, and with the structure of his family company. He changed the ending of his first name in order to reflect mainstream modern American spelling. He did not stop there though because he set to work diversifying his publications so that his customers could have access to the latest chemical knowledge, and he evolved with the times to build a strong company that would reflect German science and American corporate capitalism. The APC was unmoved. Despite letters and protests from Merck himself, just as Palmer was about to raid his company, the principal did something rather rash. He turned over 80% of Merck stock, which essentially was E. Merck's (from Darmstadt) stake in the company. It was unprecedented, and yet, it had almost no effect on the unscrupulous Palmer or his office. Merck sounded German, was based on German designs, and even though it was geographically separate from E. Merck in Darmstadt, it was still a subsidiary at this point. By 1919, the APC made a fateful decision about the possibility of Merck reacquiring their shares; they chose to stage a public auction. We are not sure what the expression on George Merck's face was like when he learned this news, but we know he started to make inquiries immediately among the largest investment firms concerning a loan [11]. An alkaloid empire with all its secrets was on the line; larger debates remained about the state of political economy, and the future of the Merck name was all in jeopardy.

A feeding frenzy was about to ensue if Merck did not defend itself. Luckily, due to their contacts and impressive resume of fine chemicals they produced, the company received financial backing from a couple of key sources, namely Lehmann Brothers and Goldman Sachs. These financial endorsements spoke to Merck's credibility and the recognition by investors of the power of German-designed chemical firms that had incorporated the very best business practices that America had to offer. Now that he was staked, George Merck put on a suit and waited for the auction to begin. A mountain of morphine and a rolodex of connections across a broad distribution network, plus all the technology of Merck's laboratories and its chemical staffs, were enticing to five bidders. We are unsure if a caller was involved, but Monsanto, a St. Louis agra-chemical company founded by John Francis Queeny back in 1901, started with an opening bid of 2.4 million [11]. He probably looked at Merck like one of those cartoon characters that is ravenously hungry and sees only a giant ham hock when it looks at its friends. With thirty years in the pharmaceutical and patent medicine business, Queeny made Monsanto (which was named after his wife's maiden name) into a power player through the sale of saccharine, which he successfully pawned to the Meyer Brothers Drug Company. He was up against it in this particular auction. Merck was not going to let the company he had painstakingly built from scratch back in 1891 be bought wholesale. No huckster from St. Louis was going to stand in his way. In 25 short minutes, which to Merck must have seemed like days, he won. With a final bid of 3.75 million dollars his creditors would send that sum to the APC and the Chemical Foundation headed by Garvan [46]. It must have been deeply satisfying, yet unnerving, that someone who had weathered such complicated legalities to become American and who in good faith handed shares over so willingly would be made to pay for his own company. Merck's evolution as a titan in pharmaceutical development and design was only beginning because rough shoals were ahead.

The company survived the wrath of the APC and advanced markedly for the first half of the 1920s. George Merck passed away in 1926, and his son assumed the role of President of the newly christened George Wilhelm, Merck & Company, which continued to dominate other competitors in alkaloid production and sales to the tune of 6.1 million a year. It seemed as though the laboratory and its practices were safe, but they were not. In order to finance the purchase of the company at the auction, Merck had mountains of debt that it owed to both its major creditors; in short, it was sinking and needed help. Fortuitously, Merck & Co. found just what they needed from an unlikely source that in the past had been a mainstay in chemical manufacturing based on the old model of family ownership. The Rosengarten brothers of Philadelphia, who had merged to form Powers-Weightman-Rosengarten back in 1905, swooped in and provided just what Merck needed. Seeking to get out of the daily operations involved in large-scale chemical production that was burgeoning in the 1920s now that competition with Germany was much more level, the Rosengarten clan only required board representation, and had no intention of dipping their hands into the affairs of the laboratory [11, 45].[25] In a twist of fate, they actually wanted to merge with Pfizer, but the Coolidge Administration ironically stated that it would violate anti-trust laws. So, instead they went with Merck, and that decision (the first of two mergers for them ahead) would reverberate throughout the rest of the century. Overnight, Merck production more than doubled and sales by 1928, on the eve of the Global Great Depression had never been higher, at over $13 million [11].

The new Merck & Co. (later known as MSD) in America provides us with an example of a Big Pharma chemical company that was transcendent. Considering what transpired for them in the wake of the Great War, it was the creation of a new type of chemical heterotopia, unseen in North America before. Family companies, like Rosengarten, that developed chemicals primarily for local outlets, were a relic of a bygone age. Now the emergence of large, science-based chemical industries, like IG Farben in Germany and Du Pont (they opened a pharmaceutical branch called the Haskell Lab of Industrial Toxicology in Newark, Delaware in 1935) in the Mid Atlantic, were the future where corporate capitalism was now headed. Merck & Co. was first, but not alone; by the 1930s, they were joined by the progressive Abbott Laboratories in Chicago, Illinois that built a state-of-the-art facility based on the designs of the famed architect, Raymond Loewy; then, the Squibb Institute for Medical Research in New Brunswick, New Jersey; and then by the Lilly Research Facility in Indianapolis, Indiana (wing of the nineteenth century firm Elli Lilly) [47]. Internationalism was firmly implanted in America (much to the chagrin of the isolationists), as it had officially arrived to spur development of those major industrial players that had the margins to finance and invest in future research. Hoover's cooperative and associative state was uneasy, but becoming fueled in new ways that were just the beginning. Financed by powerful brokerages and backed by a continually evolving biochemistry, markets of competition took the lead. Germany, for its part, was not out of the game by a longshot, even though they would experience, first a financial collapse, and then, the rise of an authoritarian regime under Adolf Hitler. In America, despite progressive legislative actions that supposedly limited the production of

---

[25]The Rosengarten brothers were Adolf, Frederick, George, and Joseph, respectively. Each enjoyed sport more than making chemicals at this point in their careers. The 1920s was marked as a time where golf, baseball, hunting, fishing, horse racing, and boxing all saw increases in activity, as those with means enjoyed rising standards of leisure.

opiates (now called narcotics), business for firms like Merck & Co. was never better. They had survived persecution at the hands of the APC, and had found a means by which to turn debt into cash for chemicals in an ironic reverse of the American Dream. Emboldened, the company now became an example of how capital when paired with science could reinvest in new laboratories which would parallel the most inventive in Europe. With a high protective tariff in place, continued debates over how clinical medicine and biochemistry could work together, a finely-tuned laboratory ready for duty, and the possibility at hand for the creation of a fresh set of both partial and fully synthetic alkaloids on the horizon, the sky was the limit [48]. Neither a Global Great Depression, nor a catastrophic World War could stop the continued and eventual rise of the opioid; in fact, those events would assist considerably.

3.2.3 War as Catalyst

During the 1920s and into the 1930s, if we can offer some generalizations, there was a sense in Western nations that wars were terribly destructive in the modern sense, but they provided opportunities for advancement, especially when pertaining to science, technology, and especially, medicine. The First World War proved that killing soldiers could become more efficient with the advent of the meat grinder along the Western Front. Concurrently, healing through the delivery of opiates like morphine, continued through a series of medical tent centers, as army life behind the lines became more and more streamlined. Like a factory floor, if rotations and drugs could be administered in sequence, then soldiers could return to the trenches and continue to fight. By 1918, with a global flu underway that destroyed young men and women as the war was coming to a conclusion, the next decade saw rising hopes that healthcare could develop a series of pain relieving treatment approaches that would incorporate the very best that the new industrial laboratory heterotopias had to offer. Before 1939, with the rise of socialistic and authoritarian regimes that were cut from similar political ideologies, some aspects of life became more regimented, like government programs, and others became more open, like options for medical treatment [39]. Though healthcare delivery and fighting pain did become more professionalized in America, it also was subject to heavier regulation after the FDA was re-empowered after 1938. Also, as we have seen within the chemistry laboratory, whether found in an industrial setting, within academic departments, or in newfangled hospitals the prescribing of pain medications continued to evolve. In the period leading up to the Second World War, research into the understanding of opiates (especially morphine as we will discuss in Sect. 3.3), and how they functioned inside the human body would play an important part in the science and chemistry of fighting pain [49]. Thus, the conflict, the killing, and the healing of pain took a major step forward as synthetic medicine and its deployment across continents hurtled forward at a breakneck pace.

By 1920, understanding more completely the long-term effects of drug use and more specifically abuse, was still far off in the future. Conceptions by members of the AMA, by legislators, and anyone else that believed in the tenets of a therapeutic reformation continued to make slow and plodding progress. At the start of the decade

Americans knew that heroin, for instance, was addictive, but despite the Harrison Act, it was still legal. We have to remember that it was only two decades before that heroin was introduced to America by Dreser's lab at Bayer. Loads of drug manufacturing firms incorporated the crystallized substance into their own production lines. To us it sounds insane based on hindsight, but heroin was viewed by many groups, including members of the pharmacological community, as an antidote for society's ills at the turn of the century. In fact, the Saint James Society, an American-based philanthropic group even mounted their own campaign to supply free samples of heroin to alcoholics.[26] They had the idea that using the mail service would be the best possible means to effectively disseminate their assistance. They had no concept that they might be colluding in the first example in modern history of a heroin drug ring via the postal service [50]. What this foreshadowed was a future connectivity between drug-makers, health providers, and marketing sales representatives who promoted the delivery of individual pharmaceuticals which were specifically developed and used to fight addictions that people had to other ones.

As the 1920 s roared in, the United States Treasury Department increased its efforts to combat abuse and what they surmised was an underground set of drug cartels that they believed were operating inside the country. With a new Narcotics Division, the federal government began to wage war against a new generation of cooks in chemical laboratories. It would have mixed results. On the one hand, it was given expanded powers when heroin sales were officially stopped with the passage of the Heroin Act in 1924, making the importation, manufacture and possession, illegal. Finally, the government opened its eyes to the growing rates of addiction that Bayer had heard about before 1917. This particular version did what the Harrison Act should have, namely outlawing even its medicinal use. The new wing of the Treasury, which would add organized crime to its targets like the famous wars against Capone in Chicago, constituted the first federal drug agency devoted to enforcing that ban on all legal narcotics sales. The other half of the story was, as with Prohibition, once legal venues to purchase heroin were cutoff, addicts were forced to buy from illegal street dealers [51]. Thus, the influence of the pavement, the back alley, and all of the subversive activities that went with it would grow stronger and more sophisticated, especially after 1945.

Marching into the 1930s, the globe plunged into a set of depressions that would have a reverberating effect on everyone involved; opiate use continued to be assessed and reassessed as alkaloids were continually scrutinized. As discussed previously, Merck & Co. continued its march towards dominating the production of semi-synthetic pain relievers that were still legal to prescribe. Morphine, strangely enough, continued to be monitored, but not banned. It still denoted images of dens run by illegal Chinese businesses and dispensers of quack medicines. Despite the best efforts to quash them by the federal and state governments they all were still present

---

[26]There is a significant discrepancy in the historical record and among scholars, as to whether this Society or for that matter any group attempted to use heroin to fight those that were afflicted with morphine addictions. As of the time that this publication went to press there has never been any incontrovertible evidence to the contrary concerning this point.

much to the distaste of the medical field. For the first part of the decade, doctors earned more and more confidence from the people, and their march to ascendancy over their pharmacological colleagues was getting even more expansive with each year. Doctor's offices and hospital districts were now found as architectural spaces located in close proximity to one another, especially in the urban built environment. As the New Deal funneled money into building projects throughout the country in a vain attempt to get the economy moving again, these complexes mirrored the universities which were training the next generation of medical personnel through expanded coursework and training [19, 39].

Laboratories associated with clinical developments such as these were becoming even more prevalent during this era. Since medical schools were getting more and more diverse, they were beginning to regularly draw from biochemistry programs which were gaining ground amid the continued struggle to establish themselves as a branch of a legitimate form of organic chemistry. It had taken decades, and the field had come a long way from the days when it occupied a small portion of the Pharmacological Department at Chandler's Cadillac Ranch at Lehigh. Applied sciences, as a division at universities across America, was finally getting its due as those connected to chemistry were realizing what Liebig had advocated all those years ago; namely, that an intersection of beliefs and ideas when examined from the chemical, medical, pharmacological, economic, and social structures could enhance any pursuit of any problem.

Despite all the consolidations and reformatting of business objectives and plans during the 1920s, the issue in America remained how to connect industrial fine chemical manufacturing with the needs of the therapeutic clinic. Opiates were still technically legal, but new forms of alkaloids were only scrutinized after their release when they were finally tested. If they failed to protect the public, only then were they pulled from the shelves. Something about this particular approach was severely lacking and just plain backward. Like the proverbial cart before the horse, it was a totally reactive strategy, if you could even deem it as such. Unfortunately, what was needed was a massive calamitous event which could shake the foundations of the medical and chemical communities, and spark action in a lethargic Congress. The choices made by the chemical laboratory at the S. E. Massengill Company in Bristol, Tennessee present an example of not only the misuse of chemical knowledge by the ownership, but also the professional staff. Clearly, though the failed response by the federal government throughout the first number of decades of the twentieth century is significant, what also resonates are the events surrounding Massengill that present a case study that will be applicable as the development of the modern opioid unfolds at mid-century [23].

It all started with something not directly connected with alkaloids. The development of a drug called sulfanilamide first originated in Austria during the Ehrlich period where laboratories were searching for what would become known in the future as antibiotics. There, a doctoral student named Paul Gelmo as part of his dissertation first prepared the new compound in 1908. After it was patented a year later, it was not until the next generation of chemists at the powerful laboratory at the Pasteur Institute in the 1930s took up the chase. They began to explore its chemotherapeutic abilities as an agent that could slice through even the toughest bacteria [19]. Likewise, to rival the product German firms attempted to recreate it as well. Eventually, in the international market of trade for drugs, which was still robust despite the global Great Depression, sulfanilamide made its way into the distribution chemical networks of the United States. It was unpackaged one day on the dock at the S. E. Massengill Company in Bristol, Tennessee. The company was in business for over thirty years since the founder, Dr. Samuel Massengill, had decided to open a drug manufacturing business instead of using the medical degree from the University of Nashville he earned to enter active practice. Management was hoping for a hit since sales were good, but competition was stiff out there. Samples were handed out by eager sales-man that received instant feedback from patients and doctors. What they heard was simple, the new drug was a miracle, but it tasted horrible. Something had to be done.

A chemical fork in the road was reached by the chief chemist at Massengill, Harold Watkins. He had heard the calls from management and knew what he had to do. In order to improve the taste of sulfanilamide he needed to be able to dissolve it so the raspberry flavoring (we are not sure why they chose this) could take full effect. In charge of the laboratory, he made the command decision to use diethylene glycol. There was one major step he overlooked, and it would be lethal. He did not order a toxicity test on any animals, thus overlooking the fact that the agent was used in antifreeze. Studies existed and journals had concluded how toxic it truly was as a chemical. Had Watkins consulted any of them, he would have learned this. Instead his slipshod approach led to this chemical heterotopia churning out 240 gallons of the newly created Elixir Sulfanilamide, as it was dubbed, in 663 shipments that went to branch offices in Tulsa, Kansas City, San Francisco, and New York. Within a few weeks the company began to hear from frantic doctors that something was very wrong. Patients, especially parents of children, were reporting sickness after taking a dram that came from both prescriptions and over-the-counter from pharmacies, and then the death toll began to rise. FDA investigators took weeks to fan out and recover roughly 90% of the solvent [19]. In the meantime, Massengill spent most of that time hunkering down behind their Bristol walls, and was primarily unresponsive to claims that they had anything to do with the situation. Once it was all over 107 individuals lost their lives, which did not include the chief chemist, Harold Watkins, who could not bear the poor choice he made, so he took his own life. What ensued during this event was a crisis in drug-making, policy construction, and a warning about the power of chemistry and marketing. Clearly something had to be done, but would an act of Congress or another federal agency truly prevent future pharmaceutical disasters?

The problems that Massengill unearthed and the federal response were significant. First, the existing legislation was flawed. The selling of lethal drugs was not illegal, only detrimental to a company's chemical reputation. The Bristol firm was only assessed a $26,000 fine, and it was business-as-usual thereafter. There was nothing in the law that strictly stated that what happened next was cause to send Massengill ownership or its chemists to prison. All the punch that the Pure Food and Drug Act had given the FDA was the ability to investigate misbranding [19]. The only piece of evidence that could assist with prosecution was the fact that Massengill's brain trust had chosen to put the word *elixir* in front of the word sulfanilamide, which according to the *USP* meant that it contained alcohol; of course, there was none to be found.

Death, especially when children were the victims, at the hands of an unscrupulous company shocked Washington into action. Yet, even that was not immediate. When politics are involved there are always interests above all else. The pharmaceutical lobby, which had grown in strength during the expansion of bureaucracy during the New Deal, wanted the power to knight new drugs to remain as the status quo, and for advertising oversight to rest with the Federal Trade Commission (FTC), which was sympathetic to their abilities to disseminate what they wanted. A fight ensued as FDR's Agriculture Undersecretary Rexford Tugwell, the future governor of Puerto Rico, sought not a flashy FDA, but one with teeth. Finally in 1938, he got his wish with the passage of the Food, Drug and Cosmetic Act [23]. Now laboratories would have to be conscious of their research, their choices, and how they were influenced by sales and marketing. Drugs at this time needed to be proven safe to be sold, but did the law have the ability to morph with the times? Only time would tell. Many medicines that were derived from opioids and already being sold such as codeine and morphine were still allowed to be used by physicians despite the passage. The whole basis for the law itself, what consumers were so angry about, was not submitted until the last minute. This portion of the bill, which forbade the sale of any new drug unless the Secretary of Agriculture (through the FDA of course) found it to be safe, might not have occurred. In the end, the government would tell the public what drugs it could and could not buy [11]. Though drug oversight went to them, it was the pharmaceutical lobby groups that won the regulation battle of advertising that was left to the ingenuous hands of the FTC.

The sulfanilamide syndrome and the subsequent backroom deals from lobbyists representing those powerful trade groups were indicative of the complexities surrounding issues such as the marketing and sale of pharmaceuticals. Up until this point they had occupied a sort of nebulous place in the chemical industry since so much of the reliance on new items were contingent upon the influence of German labs. Now, right before the Second World War, with an FDA emboldened with oversight, it seemed that patients would have the protection they needed from quack medicines, rogue doctors, and laboratories that were unable to properly lead experimental studies. Perhaps most of all, the Massengill disaster uncovered the tenuous role played by the laboratory.[27] The AMA used the event to downplay the deaths because as they argued, quack medicines killed many more than an isolated incident like the one in Tennessee. Who was then really to blame; administrators, sales representatives, local and state officials, or federal government entities like United States Customs, the FDA, or even the Congress? Or were all of them in some form complicit? In this instance, was it not the chemists, who missed the toxicity levels in diethylene glycol? After all, they were the ones with the technical skills, the knowledge of the periodic table, and certainly, they would be the ones to spot a product that was more flavored elixir than a miracle medicine. It rested with them. Again and again, the pressure from the industrial chemical market, the variance

---

[27]The Massengill Company continued to operate as a family-owned pharmaceutical firm until it was acquired in 1971 by Beecham. It later merged into SmithKline Beecham in 1989, and since 2000, merged into the giant conglomerate GlaxoSmithKline.

in the types of laboratories, and the training of their personnel continued to be the aspect of this story that was unregulated. Additionally, it was about to get better and worse in several regards after the Second World War.

Before the FDA could become acclimated to a new role as enforcer, the world changed. The Third Reich rolled across Europe, while the Empire of the Sun hopped all the way to Australia. For four years these two newcomers, who unified from within back in the 1870s, came to dominate not only the ground and sky in front of them, but they continued to develop new drugs that could be used to enhance their conception as Aryans in the West and Asia for Asians in the East. In a twist of Orwellian irony, the Nazis leadership feigned morality by outlawing highly addictive alkaloids, and then summarily proceeded to abuse them behind closed doors [52]. The effects of what those German laboratories produced would have a profound influence on the American market (as will be discussed below in Sect. 3.3), as they had before. In the meantime, a world war was being waged up and down deserts, along beaches and hedgerows, and in towns and cities in massive operations. An air war developed alongside tank campaigns that were totally motorized, and there was no break during the wintertime. From 1931 to 1945, mounting casualty and kill ratios of both soldiers and civilians were meteorically rising. Saving privates to housewives became a daily struggle as almost no one on the planet went untouched by the events of the Second World War. Morphine by the gallons was created and injected like never before, and pharmaceutical firms vied for the opportunity to increase business and diversify their economic structures. Wars, as previously discussed, accelerated cooperative attempts at science and medicine because it made sense to pool scarce resources. Likewise, to find efficient and effective ways of keeping people alive was never ending in order to stave off annihilation. The United States converted its bureaucratic attack from resuscitating the economy to throwing its full weight behind networks that would link laboratories, disciplines, and other groups together in order to streamline solutions. President Franklin Roosevelt funneled research money into the newly created Office of Scientific Research and Development (OSRD), which made the old Bureau of Chemistry's budget seem minuscule. Medicine also received a major boost from the First World War entity of the National Research Council's Division of Medical Sciences, as they turned university laboratories into a set of spaces tasked with specific goals [53]. Some tackled penicillin production and a new era of bacteriology awakened, while others went after gonorrhea, a major concern when the doughboys were around back in 1917. Within the globally-minded OSRD their chief Vannevar Bush, created the Committee for Medical Research (CMR) to oversee government planning and the all-important contracting for medical research.[28] The NRC was not a governmental operation, so ever-the-politician, Bush cleverly used the CMR, which was intended to facilitate and support recommendations made by the civilian-based council, as a means to run interference and stay above board. These contracts were

---

[28] Vannevar Bush (1890–1974) was an American engineer, inventor, and a science administrator during the Second World War who headed the U.S. Office of Scientific Research and Development (OSRD), which included the early administration of the Manhattan Project. He was chiefly responsible for the movement that led to the creation of the National Science Foundation, and work in early computers for Raytheon.

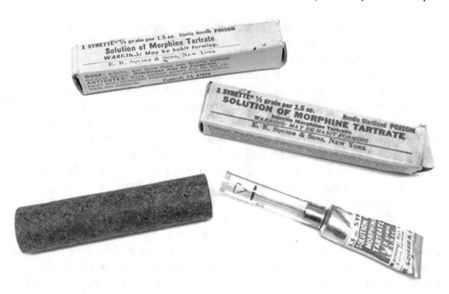

Fig. 3.10 Morphine Tartrate 1 Tube (hard fiberboard tube was used to protect the syrette in certain types of First-Aid Kits notably the Parachute First-Aid Packet). Photo courtesy of the Author

lucrative ventures and in wartime there were sure to be those in need. The production of new and old pharmaceuticals, especially like morphine derivatives, as it did during the Civil War, became an important treatment on and off the battlefield. The Squibb Company had the inside track when they developed a miniature way to deploy the pain relieving substance [19]. Once again, design met the laboratory to produce a crucial, yet at times imprecise, alkaloid-delivering device—the syrette (Fig. 3.10).

For a few years the E. R. Squibb & Sons Company in Brooklyn, New York, one of the major players in pharmaceuticals alongside MSD, feverishly worked on a new concept. They wanted to find a means to place the proper amount of morphine into a delivery device that could be used anywhere. Perhaps the deaths in Manchuria or across Europe that were continuing to build, told them it would be a good time to invest in such a venture, but for whatever the reason, their timing was impeccable. In April 1939, about five months before the Nazi *Blitzkrieg* turned Poland into a speedbump, they submitted a patent request for a *Hypodermic Unit*, which was hermetically sealed, had a collapsible metal body, and contained a one-half grain of morphine tartrate [54]. For the time period, it was a rather ingenious design because into the housing of the device a sturdy hypodermic needle was attached. Inside of this was a wire loop, which could pierce an interior seal and allow the liquid to travel down the needle, past the epidermis, and into a person's blood stream. The tube's needle was protected by a cover which would keep it from being bent in transit. Dubbed a syrette, this implied a small syringe that was easy to use and even to say. It became one of Squibb's bread and butter products during World War II. Bush's

CMR was ecstatic; now, the medics on the battlefield could deliver a safe means of easing pain. Still, there would be issues.

After America's entry into the global war in late 1941, Squibb went into action. They needed assistance to procure the mass amounts of alkaloids they would need, and of course, the materials for the syrettes themselves. Production was accomplished by both men and women, and scaling was strategically constructed much more readily than in previous decades. Americans had worked hard to perfect the art of making war. Tests of the device related that the full effects of the morphine would occur twenty to thirty minutes after the injection was applied. Initially, only combat medics were given them, but by 1944 a new breed of soldiers was given them since their jobs were extremely dangerous, and they could easily end up isolated and alone. The paratrooper divisions of the 82nd and 101st Airborne were scattered all over Northern France that early June [54]. Each of them was issued a syrette in a kit, along with a host of other weighty equipment, that included other devices like a metal cricket for signaling others in the dark. Considered precious since they could keep a person alive in extreme conditions, they became essential during Operation Market Garden and the Battle of the Bulge. The major issue for medical personnel was keeping track of how many were given to a soldier at a time. Especially in cold weather, the morphine did not travel through the bloodstream as quickly as it did when it was temperate. So, men sometimes were shot up with several at a time by numerous medics or their pals, since it was not known how many they were given prior. To guard against this practice, which would delay surgery and sometimes lead to death, trained personnel were instructed to pin used ones to the patient's clothing or would mark helmets with the customary capital M [54]. This last development was an example of functionality driving form by way of providing the most instantaneous pain relief possible. In combat conditions, new practices opened new avenues. If alkaloids could be delivered safely in the middle of a global war, they could be used and abused anywhere.

After the development, control, and use in the field of this new morphine delivery system the government contract with Squibb mandated that each kit be marked with a specific disclaimer and narcotic label. In the wake of Massengill, if the OSRD was seen sanctioning narcotic use without some justification, it would be political suicide. History dictated that the sticky substance could easily be abused. Thus, there was a continued desire to oversee what the laboratory had wrought; this in particular was a running theme after 1945 as well. Yet, in the middle of chaos, which by the end would see in the inauguration of the nuclear age, it seemed an innocuous bureaucratic relic when juxtaposed against the backdrop of a mushroom cloud. Once again, the laboratory had walked the fine line between prosecuting healing, sowing addiction, and making money. Still, the FDA and the federal government had no other option. In the post war era, where a complex set of networks dictated a constant revolving door of products that posed threats to patients, it seemed one more step towards the laboratory being further constrained by the arms of the government. As the Reich was defeated, instead of lasting for one thousand years, America reaped the industrial benefits not only from the hegemonic power that was earned, but from the chemical playing field they now commanded. As opiates morphed into opioids, war, science, government bureaucracies, health providers, and laboratory heterotopias seared in their minds the possibility of new pain-free frontiers as consumers took precedence over patients; or so the large pharmaceutical companies hoped.

3.3 Designing Numbness: Not Your Father's Opiates

Whenever a man denounces the mind, it is because his goal is
of a nature the mind would not permit him to confess.
—Ayn Rand, *Atlas Shrugged* c. 1957 [55]
...the spectacle of people following current custom for lack of will or
imagination to do anything else is hardly a new failing...
—William H. Whyte, *The Organization Man* c. 1956 [56]

3.3.1 Lab Coats of Grey Flannel: Dawn of the Opioid

In the late 1950s, around the same time as President Dwight Eisenhower won re-election after pounding rival Adlai Stevenson for a second go as he had in 1952, two books were published that helped to define the age. One fiction, and the other, decidedly not; but, both spoke about how organizations could corrupt, and yet, also promise inventiveness and individualism. Ayn Rand's *Atlas Shrugged* expressed how the mind, when free, was the ultimate expression of one's own self; taken away by structures, it would cease to exist. The novel's hero John Galt personified this by proclaiming that, *he would stop the motor of the world*, if the looters stood in the way. A celebration of the triumph of corporate capitalism, rather than as a social critique, William H. Whyte's *The Organizational Man* was a trumpet for plans and people. For him, the gray flannel suit was a uniform for progress, as a definition for order. Both books forced readers to contemplate their own existences and presented what seemed as practical applications, when applied to the creation of the modern opioid in the chemistry laboratory.

In less than four decades, the American pharmaceutical industry was transformed as a leader in chemical production; no longer the handmaiden of German laboratory designs or influenced by the cooks who plied their trade with unpatented medicines laced with opiates, now business-minded Americans were poised for a great bio-chemical leap forward. Depending on a mix of science and by emphasizing new methods of drug design that jettisoned trial and error (emphasizing the randomized experiment), companies such as MSD, Parke-Davis, Squibb, Eli Lilly, and many others hitched their wagons to the possibilities produced in the machinations of what was becoming the professional chemistry laboratory. It was not an overnight transformation. As we have seen, they were forced to look outside their industri-alized world which allowed a host of sources to begin to influence them; some of the brightest were found in the university, while others still were housed in heterotopias that scholars have tended to overlook. In the wake of an elongated period of war and depression, Whyte's corporate capitalism merged into an intricate network composed of the coordinated industry, the matriculations of the university, and was overseen by the daily functions of a government who needled and rankled the Ayn Rands of the world (there was really only one) [57]. It was a triad of power, wealth, influence, and lastly, the world of the chemist. Once the vocation where male authority reigned now joined increasingly by women, chemists were becoming members of an all-embracing system that balanced science and profit-seeking. Chemists were the engines from which new discoveries in drug-making originated; and, as will be argued here, were absolutely integral, even though industrial pharmaceutical boards of companies would be reluctant to admit it, as the most important portions of the system. Yet, this also would also leave them open to precipitous decision-making.

The clinical laboratory, that had received such an accelerant before the Second World War was also central, but it was not considered part of this movement to develop new opioids. On the outside looking in, this laboratory struggled to keep pace, as the next generation of biochemists from universities migrated to the industrial and governmental laboratories. What was gained during the 1930s was now lost, as patients were viewed as full-fledged consumers, rather than as complete human beings. For instance, between 1939 and 1959, drug sales rose from $300 million to $2.3 billion, with prescription drugs accounting for all but approximately $4 million of the increase [58].[29] The inauguration of the synthetic opioid age was moving along at such a pace, that opiates, those alkaloids discovered so long ago back in Germany, would be a thing of the past. Yet, during the 1950s just how did American Big Pharma develop what would become the opioid? The answer lies in the crafting of a new approach—the drug design. That, coupled with a new apprenticeship system within the laboratory, would become the process which had a major impact on the future of pain medication and the ways in which the next generation of chemists would be trained. The hermeneutics of corporate chemistry would never be the same.[30]

---

[29]This expansion was also assisted when the Congress in 1951 passed the Durham Humphrey Amendments to the FDCA, which created a statutory definition of prescription drugs to include those that "because of [their] toxicity or other potentiality for harmful effect, or the method of [their] use, or the collateral measures necessary to [their] use, [they are] not safe for use except under the supervision of a practitioner licensed by law to administer such drug[s]" [58].

[30]See Michael Lewis. *The Undoing Project: a friendship that changed our minds* (New York: W.W. Norton and Company, 2016). Lewis examines the work of two Israeli psychologists, Amos Tversky and Daniel Kahneman, who influenced the development of behavioral economics by arguing why we should not trust human intuition. Specifically, Lewis, in "Going Viral" outlines the ways in which evidence-based research in medicine influenced the decision-making process for diagnosing patients. This is particularly applicable to the studies by pharmaceutical firms that were carried out with opioids to this day. While the trial is not the subject of this volume, it is an important extension of the chemistry laboratory, especially under FDA approval (Chap. 8) pages 212–237.

3.3.2   Risk-Averse Chemistry: Nazi Drugs and Massengill Deaths

At the conclusion of World War II a generation came home. Elated at their ability for survival and blessed to arrive in one piece, men and women had to rapidly adjust to demobilization. For women, who did more than just operate rivet guns, the future appeared bright, but it was to be a stillborn existence. At its conclusion, advancement was fast-paced and full of opportunity. Women, who were previously disqualified from a host of positions, were thrown into new positions. They filled them more than adequately [59]. Companies, like Squibb, had thrown their doors open to staffs that did everything from making alkaloids to the stamping of syrette kits. Once men returned, many, especially in the laboratories, resumed their old jobs and tried to pick up where they left off.

The defeat of the Nazis at the hands of those Allies who hit those beaches, braved subzero temperatures, and ducked in and out of hedgerows in country created a much different situation in peace than they did in 1918. Bombing campaigns were just as lethal as atomic explosions during the conflict, and the fire that rained down on industrial centers, like Dresden in February 1945, was absolutely devastating. Broken from stem to stern, Germany emerged a partitioned nation that was parceled out among the victors. Pharmaceutical giants survived, but the American economy emerged from the wreckage in charge of a vast set of resources and in firm control of the West German state. Interestingly enough, the zone administered by the United States included the Valley of the Chemicals, the home of E. Merck. Certainly, this acquisition was not lost on the government in Washington as they surveyed their newly conquered territory. The end of the Axis world in 1945 presented a unique set of opportunities. Unlike in 1919, the Americans were occupiers, and now had access to a vast set of resources at their command [60]. As previously discussed, pharmaceutical production was owned by the Germans, especially when it came to alkaloid scaling and production. Now, the Americas had caught up, but it was not necessarily due to their grasp of biochemistry. History dealt them a royal flush. They would play it, but German synthetic opiates were still important and had a profound effect on the American chemical market.

In the 1930s, the Nazis built a powerful war machine, along with a chemical conglomerate that was second to none. Carl Duisberg's IG Farben had engineered a set of chemicals for public consumption that was changing the face of pain relief and therapy with the development of two drugs, namely methadone (1937) and pethidine (1939). To begin, methadone was the first fully synthetic opioid analgesic (Fig. 3.11). Interestingly enough, it was actually discovered by accident. German chemists were looking for a synthetic substitute for a natural acetylcholine blocker known as atropine. However, upon further testing it was discovered the new compound was as effective at pain relief as morphine [52]. Developed by the imminent chemist, Gustav Ehrhart and assisted by his superior, Max Bockmühl, it was a revelation in pharmaceutical design. Ehrhart, who would go on to lead the Hoechst AG after 1945, conceived of the drug as a new painkiller. Working in the best laborato-

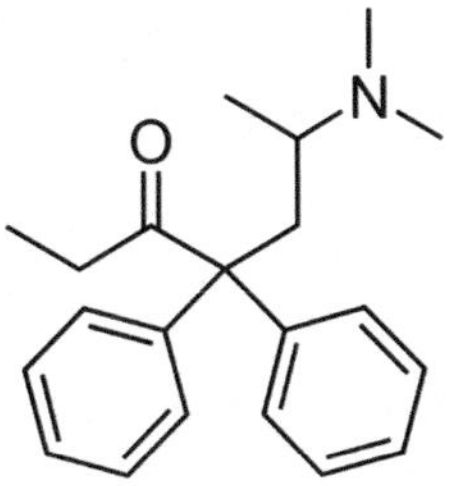

Fig. 3.11 Methadone (1937)

ries that Europe had to offer and with a wealth of experience, in the winter of 1937 and into 1938, the duo began to synthesize over 300 compounds. These contained diphenylmethane as a central structural element. By 1939 they gave the compound (±)-6-dimethylamino-4, 4-diphenylheptan-3-one the development code VA 10820. Animal experiments, long the hallmark since the days of Sertürner and his dog, these chemists found that it had a five to ten-fold stronger analgesic effect than any other synthetic opiate. A new age of opioids, though not named so, was upon them. In mid-1941, amid the chaos of France's fall and the Battle of Britain over London, it was time for VA 10820 to officially get a new name. They called it Amidon. Once the patent was filed on September 11, 1938 for the whole substance class, Amidon did not undergo any further clinical trials. However, due to the dispossession of IG Farben's secrets, VA 10820 came to the United States by 1949, where it was given the international nonproprietary name, methadone. In the same year, Eli Lilly marketed it under the brand name Dolophine. As IG Farben reformed into Hoechst, Ehrhart marketed the drug as a potent painkiller under the brand name Polamidone. In the United States, now renamed Demerol (produced by several different companies), similar to so many other synthetic opioids, was thought to be safer and less addictive than morphine [52]. With short term use, Demerol had the same pain relieving effects as morphine, but with less sedation, and not as an intense euphoria. However, with long term use over time, users who continued to use Demerol found themselves having to up the dosage amounts to achieve the same pain relieving effect. Soon it was clear users who turned to the drug because it was less addictive discovered they were in fact addicted. America's timing was both fortuitous and fateful, as access to this new drug became readily available.

Like methadone, another powerful drug hit the German market right before Hitler sent the Luftwaffe into Poland in September 1939. Synthesized that year as a potential anticholinergic agent by the German chemist Otto Eisleb, its analgesic properties were first revealed by Otto Schaumann while at IG Farben. Pethidine was the prototype of a large family of analgesics including the pethidine 4-phenylpiperidines, the prodines, bemidones, and others, including diphenoxylate and analogues (Fig. 3.12) [52]. Like its cousin morphine, pethidine exerted its analgesic effects by acting as an agonist (meaning it produced morphine-like effects) at the μ-opioid receptor. It had structural similarities to atropine and other tropane alkaloids and might have had side effects. In addition to these opioid-like tendencies (a true -oid), it possessed a local anesthetic quality that related to its interactions with sodium ion channels. Pethidine's

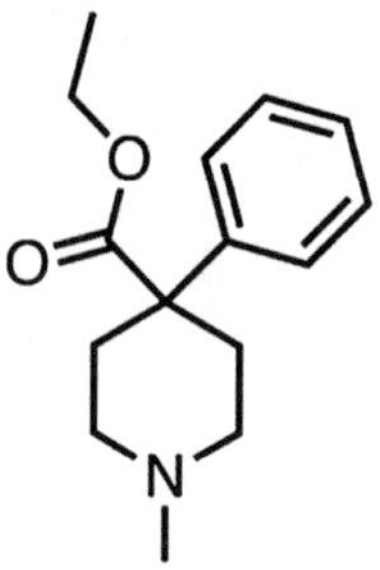

Fig. 3.12 Pethidine (1939)

apparent efficacy as an anti-spasmodic agent was analogous to its local anesthetic quality. The drug also had stimulant effects that were mediated by its inhibition of the dopamine transporter (DAT) and norepinephrine transporter (NET). Several analogs of pethidine, such as 4-fluoropethidine, were eventually synthesized and were potent inhibitors for monoamine neurotransmitters dopamine and norepinephrine via DAT and NET. It also turned out to be more lipid-soluble than its predecessors, resulting in a faster onset of action. Pethidine was shown to be less effective than morphine and later, diamorphine or hydromorphone, at easing severe pain, or that which was associated with movement or coughing. Like other future opioid drugs, the German pharmaceutical had the potential to cause physical dependence or addiction. As we will see, it would be abused more than other prescription opioids, due to its rapid action. For members of the Reich citizenry that could obtain it and then the American public, pethidine was consistently associated with euphoria, difficulty concentrating, confusion, and impaired cognitive performance when administered to patients [61].

The German synthetic opiates flooded the American markets after 1945. Doctors, journals, and associations were inundated with calls from patients hopeful that a pain revolution might offer hope. Soldiers, who had returned home maimed and debilitated by the horrors of war also waited for the FDA to move on accrediting these new pharmaceuticals as completely legal. The FDA though was skeptical and developed a risk-averse strategy that matched the period of corporate capitalism that Whyte discoursed in *The Organizational Man* [56]. In this situation with the FDA, it helped to protect patients and consumers from the onslaught of drugs that were saturating the pharmacies and hospitals. Still smarting from the effects of 1938 and the Massengill disaster, they were wary of German made products. The Nazis unpopularity of course to led to skepticism within the scientific and medical communities. Yet, the systems in place since America's entry into the Second World War in 1941 were vast and capable of handling a complicated set of requests. The FDA was no longer a small government agency that was found during Harvey Wiley's days at the Bureau of Chemistry; now, a large and powerful bureaucracy could address a full range of drugs and their effects (and it would continue to grow in strength after 1965). The age of drug design, coupled with a risk-averse approach had arrived, just as a new class of chemists was poised to redefine the chemical heterotopia like never before.

3.3.3 The Education of the Company Man (and Woman)

If you were an undergraduate interested in a career in chemistry on a college campus in America right after the Second World War you might be searching the catalog trying to decide what courses to take. You were most definitely a white male, just as it had been for generations, and you probably came from some means. Yet, after 1945 things were changing beyond taking Chem 101 and 102. Integration of ethnic minorities was coming, and many schools would have to grapple with suits calling for co-education between the sexes in the next decade. Some smaller colleges were progressive at both of these integration attempts dating from the nineteenth century, but their laboratory programs were at times blunted by financial constraints. The G. I. Bill, which would offer so many opportunities for returning soldiers to earn a degree, would push women out of the gains they had made during the 1930s and early 40s. Still, all of this had a reverberating effect on the industrial chemistry laboratory. In order to understand the opioid, we must delve into the connection between institutional contexts and disciplinary styles that created the next generation of company men, and women, for that matter. The historian of science Charles Rosenberg aptly pointed out that disciplines ultimately shape the scholar's vocational identity [62]. All of this would influence the pharmaceutical heterotopias of the drug companies of the 1950s, and the debate over just where the teaching of biochemistry would take place presented itself front and center.

In the mid to late nineteenth and early twentieth centuries powerful European industrial giants were at times greatly influenced by social movements that weaved their way into chemistry; namely, in the German coal tar industry with the development of products such as beef extract and aspirin (1870s) and in Britain with the creation of antibiotics (1930s). In Central Europe, Justus Liebig and his students fanned out across chemistry and formed their own version of the applied sciences by using their own professional ideologies as levers to create new institutions. At the time, these took the form of industrial laboratories where professors served first as consultants, and then as hired hands. By the late 1920s in England, academic laboratories and research firms combined to fan the flames after Alexander Fleming stumbled over a petri dish after being on leave that was full of a brand-new bacteria culture [60]. Still prior to 1940, disciplines were not homogeneous and structured in the same vein, which meant a chemical chaos of sorts. Different programs adapted to different contexts and ideas. Disciplines like organic chemistry had bouts with biochemistry, which developed three distinctive styles: biological (utilitarian), bioorganic and biophysical (narrow and traditional), and clinical (broad and cutting edge). Each of these recruited its own camps in the hopes of installing department heads that would be sympathetic to their ideological underpinnings. By the 1930s, biochemistry developed into the role of pedagogical leader within the increasingly professionalized medical schools, but it was a marginalized subject within the chemical departments of America. Hardliners just did not see the benefit of appointing biochemists as assistant professors. So, if you were an undergraduate seeking to study biochemistry it would only be offered through medical school programs. Compounding this, few biologists were recruited for biochemical programs, and to top it all off, even biology departments shied away from appointing biochemists. Except for a close relationship with clinical pathways, which was rocky at times as well, biochemistry really blossomed only in the inter-disciplinary specialized research laboratories of the most progressive schools in the country [3, 19].

Everything across science began to shift in the 1930s. Hitler's refugees that possessed advanced science degrees landed into the welcome arms of the United States government who not only did not speak German, but also the language of industrial science and theory. This vast amount of knowledge would fuel all sorts of research opportunities and results; from experiments in a squash court at the University of Chicago to antibiotic investigations at fruit stands in the Midwest. A major expansion in the biomedical applied sciences resulted that was aided by new laboratory technologies that were backed, despite the Depression, by copious amounts of available funding from sources like the Rockefeller and Macy Foundations [3]. The only aspect holding biochemistry back was the specter of previous roles, values, and habits. Immediately after the war, graduate student rates in biochemistry expanded exponentially as postdoctoral fellowships supported by those powerful bureaucratic entities created under FDR blossomed. Namely, the National Institutes of Health, the National Research Council, and the numerous other private foundations jockeyed for the best brains among national and international students. Suddenly, the United States, in the wake of war-torn Europe, pressed ahead and the momentum was theirs. That new generation of postdoctoral fellows became the engine for pharmaceutical change into the 1950s.

If there was a museum of laboratories that displayed dioramas of different historical heterotopias, the one from 1950 would be of particular interest. At some point, women began to re-enter the space. We emphasize, re-enter, because there are examples from history (Madame Lavoisier to Madame Curie come to mind) where they forced their way in. They were eminently capable, but the male-centric laboratories, just as with healthcare delivery, were the order of the day. Still, times were changing, as academic institutions began to develop sister-relationships with larger and better funded research labs. Unlike colleges and universities, corporate America did not have to reflect the same bias; that is not to say they did not; it is just that the opportunity was there because of two extenuating factors. The first was the results from the global conflict of 1939 to 1945, which at first bent gender roles, then supposedly returned to pre-war traditions. However, by 1950, one in three women worked, and this had a huge effect since it showed what a workforce led by them could do. Given the opportunity they rose to the occasion and were absolutely seminal, and not just in menial positions. The second factor that was even more prominent was that women labored to earn spots alongside men in that day and age when you could gain entry based on pure scientific skill. But, the overriding fact was that laboratory doors were controlled by men. It should be noted that women in scientific societies and as parts of women's groups lobbied both the government and the field of chemistry by arguing that their capabilities were on par with their male counterparts. Several of them, such as Sigma Delta Epsilon (women's scientific association around since 1921), put forward grants-in-aid for applicants and even re-training programs for those that had dropped out to start families. The chemist JoAnne Mueller, who had a master's degree from Indiana University, conducted over 40 h of research a week on one of those grants. She did all of it from her very own basement laboratory heterotopia in her house, all the while taking care of her infant. Though calls continue for the entry of more women into the lab today, this is not a new phenomenon by any means. Biochemistry was for men, as well as women [59].

Besides the construction of a new diorama that began slowly to include women, an example of a shift from traditional research to new applied perspectives occurred at places like the University of Michigan. Following the leads of Harvard and Columbia, schools began implementing microbiology, as it was beginning to take hold [3]. Though new equipment was purchased with the intent to set up labs that would pursue this burgeoning field, at times teaching loads and rapid additions to what seemed to occur on a weekly basis, was transforming old modes and approaches. This was particularly reflected in textbook development. In an age before websites and Power Point, textbooks were the primary vehicle, besides the lecture and labs, for communicating knowledge. Multiple editions were continually issued in order to keep up with innovations in the fields. A whole generation of books appeared in the 1950s which catered to different divisions among biochemistry. One of the most popular was written by Benjamin Harrow of the College of the City of New York. His popular work with the very original title, *Textbook of Biochemistry* went through over five printings [3]. A review in the *Journal of the Chemical Education* by Dr. Abraham White, a professor at the Yale Medical School, said it best when he purported that, *each new addition of Harrow which has appeared gives indications that the author is continually approaching the production of a textbook which would become widely adopted for medical school teaching* [61]. As such, it makes sense that White would offer this narrow appraisal because his background in instruction that originated in a medical school. Yet, White and many of his other colleagues recognized what the hardline organic chemists did not; that being, when biochemists remained within the confines of medical programs they could influence a cadre of future clinical personnel that would change the face of medicine across generations. As future company men and women, their backgrounds schooled in the heterotopias of the clinic were essential building blocks for modern opioids. The next legion of organizational chemists that were bound for Whyte's world of corporate America, who were both male and female, would take part in the greatest expansion of pain relieving design recorded. The case of Dr. Small and his alkaloid-producing laboratory is particularly instructive as an example of how biochemistry spawned a new era for Big Pharma.

3.3.4 The Man Who Knew Better: The Case of Dr. Small

Lyndon Small knew how to crack one off. He knew how to properly manipulate the speed of the air through the pipe. He knew chemistry and knew how borosilicate glass had changed everything. To be a glassblower, you have to understand chemistry. Heat, timing, patience, and motion all combine to make this art a difficult process to master. Dr. Lyndon Small was a master glassblower, but he was also a master chemist

[62]. Matriculating at a time when so much was occurring in a chemists' world during the 1920s through the 1950s, Small would build a dossier that was strident in his pursuit of truth in chemistry, possibly with detriment. His trade was alkaloids and his specialty as chemist that lasted for over two decades was morphine. Dr. Small had the best of intentions, but the reverberating effects of the policies of the Committee of Drug Addiction and his own research, would set in motion a chain of events that would contribute to the development of the modern opioid and beyond [63].

Small's (1897–1957) road to becoming the director of the United States' first Drug Addiction Laboratory in the history of the country was an excellent hire and made complete sense. After growing up in Massachusetts and serving briefly in the Army Medical Corps, he built a career by inhabiting a diverse set of laboratory heterotopias that included stints at Dartmouth, Harvard, and MIT. With such a wealth of experience he actually followed the model of so many of his predecessors, he went to Germany. James Bryant Conant, Small's mentor at MIT and one of the most influential organic chemists in the country, recommended him for a one-year fellowship to the University of Munich beginning in 1926. It was there that he met and worked closely with the esteemed and humanitarian chemist, Heinrich Wieland (Fig. 3.13).[31] In the space of one year, Small had two of the titans of chemistry as his guides. A worthy student, Small utilized every possible avenue, and even adopted local customs. He brought his wife along, Marianne, who served as a life-long partner. She developed an innate chemical sense, and like Madame Lavoisier, served as Small's editor, research associate, and consigliere. Like so many chemists before him, Small was poised to gravitate to a subject that would occupy the rest of his career—alkaloids. Wieland gave him a choice of topics, and his student chose carefully [64]. For Small, pursuing synthetic opiates was part science, part passion, and bordered on obsession. His singular goal from this point in his career was to create something wholly new; a piece of chemical history that would reverberate, succor, and replace past opiates, which had torn at people's stomachs and made them into mindless idiots. Small did not display an outward social mission along the lines of Liebig, but still, he was off and running.

---

[31] James B. Conant (1893–1978) was an American chemist; known as a transformative President of Harvard University (where he advocated the use of the first SAT scores), and the first U.S. Ambassador to West Germany. Conant obtained a Ph.D. in Chemistry from Harvard in 1916. During the First World War he served in the U.S. Army where he worked on the development of poison gases. He became an assistant professor of chemistry at Harvard in 1919, and the Sheldon Emery Professor of Organic Chemistry in 1929. He researched the physical structures of natural products, particularly chlorophyll, and he was one of the first to explore the sometimes-complex relationship between chemical equilibrium and the reaction rate of chemical processes. He studied the biochemistry of oxyhemoglobin providing insight into the disease methemoglobinemia, helped to explain the structure of chlorophyll, and contributed important insights that underlie modern theories of acid-base chemistry. Heinrich Wieland (1877–1957) was a German chemist who won the 1927 Nobel Prize in Chemistry for his research into the bile acids, but he did much more for the studies in the field. Wieland earned a chemistry degree from the University Munich and afterwards worked on myriad projects that ranged from research in alkaloids and poisons associated with wild mushrooms. He did not sympathize with the Nazis, as he attempted to shield Jewish students who

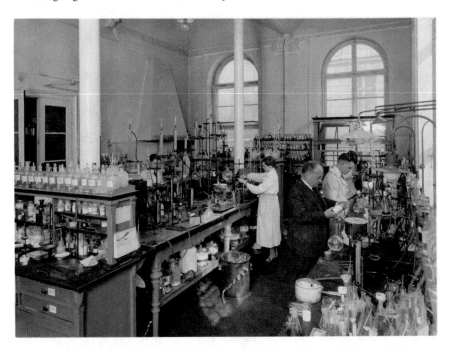

Fig. 3.13 Heinrich Wieland in the Baeyer Laboratory, Munich, Germany (notice, his progressive notions about allowing female assistants to work in his laboratory). Photo courtesy of Edgar Fahs Smith Collection, Kislak Center, University of Pennsylvania

His chemical internship complete, Small's career would reflect three major developments that were part of the period in which he was working; these included, the role of references, the direction of medicine, and the expansion of government. At the behest of Conant, Small and his family decided to return to Harvard where the part-time glassblower would take up residence in Cambridge as the master's primary assistant (for an image of Dr. Small later in his career, see Fig. 3.14). The way of the chemist in the 1920s and 1930s were no different than today. Knowledge of someone who could fill a specific post would be perfect for a rising post-doctoral student who was looking to set out on his own and begin a new career. Likewise, it would assist the one giving the reference by placing members of their cadre at significant posts. A chemical patronage was something Conant worked assiduously to cultivate, as he tended to the next generations of professionals. Since he knew Small was specializing in alkaloid research, he thought that recommending him for a research associate position at the University of Virginia would be the right fit. Small returned to the United States and with his family in tow, moved straight to Charlottesville. They arrived at an interesting time in American pain relieving history. As discussed, universities were increasingly investing in clinical medicine with the establishment of new hospital

---

were expelled after the Nuremberg Laws in an inner circle that was considered guests of the Privy Council there in Munich. He passed away the same year as his one-time student, Lyndon Small.

Fig. 3.14 Reprinted with permission from (Photograph and Biography of Lyndon Frederick Small *The Journal of Organic Chemistry* 1952 17 (1), 1–1. https://doi.org/ 10.1021/jo01135a600). Copyright (1952) American Chemical Society. [73]

LYNDON FREDERICK SMALL

districts that were directly on campus. They possessed the land, the programs, and of course, the professors. Now their medical schools would have training grounds particularly structured for clinical rounds. The communities in the college towns, like Lexington, Knoxville, Athens, Columbus, Ann Arbor, and a host of others from shore to shore, would get a full-fledged working hospital [63]. In turn, the university would get a load of willing patients seeking the best treatment possible.

When Small arrived at UVA he found a laboratory setting very different from the one he had left behind in Wieland's chemical oasis. The school was very tradition-based in its approach to education, and the making of Cavaliers for the all-male institution had changed little since the days when students rode the Lawn on horse-back, taking potshots at Jefferson's Library Clock. The Conant-Wieland student applied himself with vigor to the task of crafting his own chemical laboratory. His timing was impeccable. News of the 1929 Stock Market Crash would have slashed his healthy budget; thus, hampering his ability to procure the necessary equipment he would need. What Small was constructing was a laboratory of microanalysis, the first of its kind. His work with graduate students, connection with a chemical network of mentors, and proximity to Washington D.C. all changed his already good fortune when the Division of Medical Sciences of the National Research Council established a new committee that would study what was becoming its own field in medicine, namely, drug addiction [65]. Lyndon Small was probably the most capable and well-placed chemist to tackle this issue. He would be all in.

Fortuitously, for Small and his future assistants, the National Research Council had just launched a program to identify analgesics that had less addiction potential than morphine and its close relatives. To date, there had not been much success in crafting policies that could address society's understanding of what later would become known as substance abuse. Originally, the Committee on Drug Addictions (1921–1928) began as an organization devoted to thinking about this topic as the province of medicine. Dating from the nineteenth century, there was this hope that was transferred from doctors/health professional to Goos-Goos (those that believed in good governance) to hardline Progressives that a way could be found where substances would fight substances (which will be discussed in Chap. 4). Yet, the 1920s shifted the argument by mid-decade as the criminalization of drug-use took hold. Before this became gospel in the 1960s, the New York attorney Arthur Greenfield, with funding from the Rockefeller Foundation, spearheaded the movement towards investigating the abuse of narcotics (a word that came into fashion after the Harrison Act, meaning *numb* in Latin) [24, 66]. A political in-fight developed and those that advocated limiting drug production won out. Drug addicts, as the former law-enforcement officer turned committee member then chair Lawrence Dunham saw it, had personality defects. If you kept the drugs out of their hands, then you could lessen the impact on their own downward spiral. Out of this powerful organization came a science-based subcommittee formed by chemists who had expertise in drug-testing and production. Led by the Harvard pharmacologist, Reid Hunt, the brainchild of the Committee on Drug Addiction (1928–1938), its sole task was finding a pharmacological answer to the opiate problem. Hunt called upon Conant, who then called his greatest student, Lyndon Small to take the lead of what would become known as the Drug Addiction Laboratory [62].

As Small conceived it, he would direct the laboratory and much like Wieland, would cultivate chemists who would both experiment and take direction. He needed seasoned colleagues and staff members that would be able to understand what some perceived as a lack of personality. For Small it was just who he was, serious, focused, but open to interpretations. Once the funding began to flow in from the NRC, he reached out to Wieland and his contacts in Germany. Although the Nazi seizure of power was still a few years into the future, events in Germany were devolving with the inflation and depression crisis firmly entrenched in daily life. The October 25, 1929 issue of Science offered a short note about Small's acquisitions. It read [67]:

> The University of Virginia announces the opening of a laboratory where a series of investigations into the chemistry of alkaloids will be carried on under the auspices of the division of medical sciences of the National Research Council. Dr. Lyndon F. Small, research associate at the university, has been appointed director of the new laboratory. The present personnel includes, Professor Erich Mosettig, Vienna; Dr. Alfred Burger, Vienna; Dr. Frank L. Cohen, Northwestern University; Mr. Jakob van de Kamp, Utrecht, and Mr. Louis Eilers, Illinois.

It was the equivalent of dream squad. Obtaining professorships for his German colleagues so they could stay in the country, the laboratory now had a staff that could begin to construct a plan of attack. First, on the agenda was setting up open lines of chemical communication between Washington and their efforts at the Drug Addiction Lab. This required someone who could organize agenda items and provide a keen eye for editing all correspondence. Small's wife, Marianne, though not officially on the government payroll, was the ideal choice since she was at her husband's side since his days at MIT. Her own brand of chemistry would help everything run smoothly as complexity gave way to products. As the project progressed, all appeared smooth since research was funneled from Small's heterotopia to a pharmacological facility at the University of Michigan headed by the well-known Nathan Eddy. This

Fig. 3.15 One of Dr.
Lyndon Small's cigar boxes.
Photo Courtesy of Office of
NIH History and Stetten
Museum, National Institutes
of Health, Bethesda,
Maryland

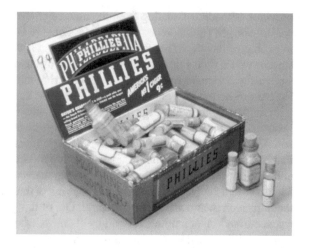

was possibly the most forward-thinking biochemical facility outside New England.
There, compounds were tested for their addictiveness and their therapeutic effec-
tiveness. Sadly, those drugs that seemed promising after numerous tests on animals
were then shipped to two facilities, Fort Leavenworth, Kansas and the Public Health
Service Narcotic Hospital just outside Lexington, Kentucky [63]. It was at both these
locations that the future of opioids would take shape as the products were tested and
retested on what was viewed as the dregs of society.

Those tests and their results did not initially yield much progress. Most of the
animals died when exposed to the intense levels of morphine-based derivatives, and
Small was flummoxed, yet patient after a draw from his trusty pipe, that results were
not coming quicker. He decided to press the pace. At this point, it fell to him, as
the chief of the Lab, to control the flow of drug samples that were coming out of
Charlottesville. Not standing on ceremony, he kept his stock of samples in a bunch
of cigar boxes with Phillies emblazoned across the lids (Fig. 3.15). He filled loads of
those receptacles with hundreds of vials.[32] One of the most cutting-edge was called
Metapon, which at the time was thought to have non-addictive forming capabilities,
and would not upset a patient's stomach [63]. The prevailing thought that Small, and
for that matter, many others ascribed to, is what one historian has called, the proof
principle; that being, analgesic properties of alkaloids could be uncoupled from the
addictive nature of the compounds. All of this thinking would become the overriding
ideology behind the opioid drug design after the Second World War, and it would
reverberate to this day with the example of Small's Desomorphine.

It all started innocently enough, as Small was looking for one of the breakthroughs.
At this point he had already built a powerful set of opiates that synthetically were
on par with any German developments. Yet, he was looking for a magic bullet that
would put an end to the tyranny of the pharmaceutical industry. Small conceived

---

[32]Some estimates from reports place the number of compounds in stock at over 500 by relating
molecular structure to pharmacological activity.

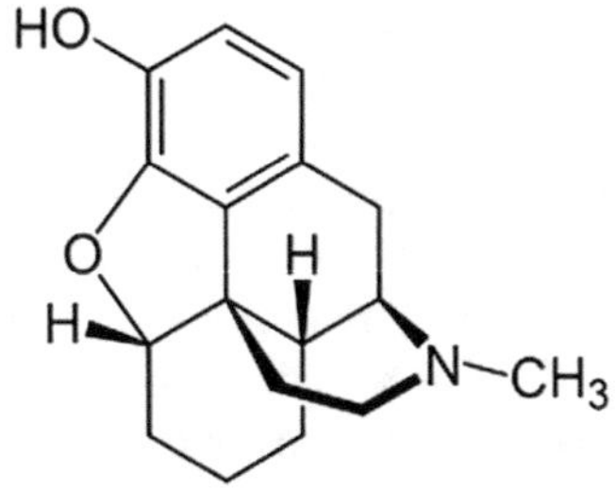

Fig. 3.16 Desomorphine
(1932)

of two ways of going about the problem that he faced. First, he would develop a series of functional groups of the morphine molecule that could be modified. He knew from past experience that this experiment might yield a break in undesirable effects, which would boost the analgesic benefits. Next, he would turn to the ring system, which would introduce functional groups of the sticky substance by placing them alongside their synthetic cousins. Small was particularly motivated when he learned of the past discovery of the safer version of cocaine that dentists everywhere would come to dispense called Novocaine. Found to be safe and not full of addictive qualities, Small, by 1932, wanted to do the same with morphine simply bypassing the urge to use again and again. What he set in motion was the American-version of what would become an opioid freight train; slow to begin, but once momentum was built, and then coupled with German developments in the 1930s, it would become a juggernaut. After several key papers, typed and edited by Marianne, the result was Desomorphine (Fig. 3.16). Studies in Michigan that were routed through Kansas and Kentucky related a drug that was eight to ten times more powerful that morphine. Promise was piqued as lower doses meant a decrease in the dreaded nausea. But, despite warnings of its addictive qualities, Washington was pleased. In fact, a patent was issued two years later and during the 1940s it was marketed in places like Switzerland under the name, Permonid [62, 63, 68]. The problem was that the stuff spoiled and lacked preservatives. That might have been a good thing since it turned out to be such a highly addictive product.

We might think that the story of Dr. Small's Desomorphine would end there, but it did not. More recently during the late 1990s, in Russia the drug's compound was reintroduced, possibly when it was discovered by the underworld or drug companies looking for inspired products, something new for the masses. It was dubbed Krokodil, which became its street name. The activities on the street had a way of infiltrating even the most fortified chemical laboratories and could form its own version of a heterotopia. When the Russian government tried to restrict the flow of opium from its Great Game rival Afghanistan (which they fought a Vietnam-style war against, and lost, in 1979), they invariably spawned what could be called popup drugs that had lain dormant from a chemical past. As street heroin skyrocketed, Krokodil became a cheap substitute. Codeine, which is readily available, got combined with paint thinner. All the accoutrements that were added could be found in the kitchen—once again, a willing substitute as a laboratory. These included matches, iodine, lighter fluid, and gasoline. This chemical soup continues to be absolutely destructive, and can easily cause death. Like crystal meth, the street name for crystal methamphetamine, the laboratory became once again a place where combinations of pharmaceuticals, when coupled with items containing high levels of toxicity could become addictive and deadly [69].

How did this happen and why did someone of Dr. Lyndon F. Small's background and capability take part in something that could be so destructive? Did he not see himself as the conductor of a synthetic train that once in motion could mow down pain, but also turn its host into a corpse? Probably not. Small remained silent about his legacy and only published in professional and governmentally sponsored journals. He received commendation after award; plus, a hearty piece of hagiography in the form of a biographical memoir from his colleague Erich Mosettig who was writing for the National Science Foundation [62]. Of course, he did not live to see what his work would become, but the question has to be posed as he was the chemical gatekeeper of the realm at UVA and later in 1950, when his group resumed work on synthetic central analgesics. Small attained marked success and the pinnacle of alkaloid research, when he succeeded Claude S. Hudson as chief of what was by then the Laboratory of Medicinal Chemistry in the National Institute of Arthritis and Metabolic Diseases. He built a resume that included the co-authorship of important textbooks, and served as chairman in the Organic Division of the American Chemical society in 1936. There is no doubt "he was a superb microanalyst and brilliant experimentalist," as one gushing historian said of him [63]. But, was Small a pawn of the policies and bureaucratic statecraft of Washington? After all, his creations, the compounds he tested in his laboratory, were funneled into a system that only believed in practical applications of solving a singular problem, finding a less addictive substitution for morphine. He should have known better. As a chemist with such power for creation, having worked where he did and soaked up what he did from the very best heritages in science, we can only conclude that he came up woefully short. Small did not become a commercial inventor of the modern opioid; rather, he chose to join government entities that were well-funded, and he reaped the benefits of a long career. However, the case of Dr. Small leaves us puzzled. If the road to the modern opioid went through his own heterotopia, if he had any inkling what he was assembling had such power and would influence the next generation of clinical research in irrevocable ways; then why was he so numb, and willing to settle for mediocrity as a professional chemist? Others would take a different tact. Those chemists of the future engaged in commercial drug designing projects and would grasp the laboratory processes of Dr. Small, which would lead to an even greater crisis of both morality and pain relief.

3.3.5 The Last Great Advocate of Reason Falls

Lexington, Kentucky after the Second World War became the home to the nation's first laboratory of addiction. More synthetic opiates, soon to be dubbed as -oids after 1963, were flowing down 421, known as the Leestown Road into the Art Deco building that is still there. By the early 1960s, opioids were being produced based on synthetic compounds and their German and American derivatives from the previous fifty years. Now clinical methods, standards, and statistics were codifying and examining, and then re-examining, the possibilities of data; a blossoming of information was upon everyone [63]. Since the founding of Small's laboratory, the influence of governmental policies, and the carrying out of drug designs on patients housed in prisons and mental hospitals, all combined to form an ideology based in seeking answers rather than asking more questions.[33] It seems that in history when this occurs, trouble looms. The world of opioids reflected age old questions of who held the high ground of chemical authority for pain relief. The last bastion of braking this slow-moving train was the clinic; that place where the latest biochemistry was taking shape in the medical schools and training hospitals of America. It would be a place of hope, but then one filled with disconsolation.

During and after the Second World War, as chemical heterotopias expanded their design capabilities with the permanent governmental installations in Washington and in Lexington, the development at Big Pharma companies everywhere continued to march on for profit. The clinical laboratory attempted to keep pace, but would it be enough into the near-future? After all it was a different world where a peace made, and nations semi-united sat exhausted, yet once again staring at the precipice with a sigh of relief. A cold wind blew, ensuring a division of the planet between East, West, and Non-Aligned that would last until almost the end of the century. The world of medicine, meanwhile, became heavily influenced by the triumph of statistics, and cooperative approaches to healthcare delivery believed that strength in numbers could outpace any illness, disease or addiction. It seemed that the art of medicine was becoming the science of detection [70]. Yet, who was driving these investigations? To put it even more simply, who was making the decisions about what drugs were good and what were, well, for lack of a better term, bad?

There is not a proper way to answer this simple question because it requires an in-depth response and further investigation. The bureaucracies of cooperative studies that Herbert Hoover had championed during the First World War were unable, during the 1950s and 1960s, to coax researchers in opioid experimentation into specific protocol for treating patients and their pain. It had been different during the 1930s when money flowed from government agencies and the Depression had weakened the business sector to the point of exhaustion. Most historians would not argue with the premise, though they might try, that the Second World War ended the downward spiral and not the New Deal. The post war era was a different time as researchers turned to the randomized experiment, which from a clinical standpoint would assist in curbing doctors from prescribing the latest and greatest treatments. An alliance formed between those therapeutic reformers, who had broken with their mentors back in the late 1920s, and the statisticians who crunched numbers and quantified usable data. Some medical associations like the AMA or publications like the *New*

---

[33]The notion of seeking answers in a rational approach to thought and processes is not germane to just the West. Certainly, it became a common issue that could be traced from the Enlightenment through the founding of the social science movements of the late nineteenth century; but, it was also part of Confucian thought in China beginning even before the Han Dynasty.

England Journal of Medicine were on board, but it was difficult for them to maintain as professional order seemed present at some points and not at all in others [19, 23].

Some opioid statisticians looked to quantify addictive behavior which still continued to prove not a disease, but the work of a deviant person that should be treated with heavy doses of synthetics and locked away as sociopaths in places like Lexington. In an effort to assist, a doctor name Clifton Himmelsbach developed a scale to measure the severity of withdrawal symptoms; a roving chart for clinicians that could serve as an onsite chemical heterotopia measuring stick. If a patient hooked on a certain morphine dosage was given a new opioid as a substitute, then it was time to watch their reactions. If the withdrawal symptoms were elevated, then the attending physician knew that that drug, fresh from the lab, was non-addicting [63]. Scales upon scales were charted and filed in a new paper revolution that would make the effects of the Harrison Act look paltry. Luckily the copy machine was not far off, since it would be necessary to chart a patient's progress over an extended time. While animals and prisoners continued to be test subjects for these drugs, over two decades of work with actual patients in hospitals yielded results that led clinicians and researchers to understand whether new compounds acted as agonists (which produced morphine-like effects) or as antagonists (which reversed effects of any addictive source). Still, there was little consensus as the 1960s unfolded about the chemical high ground for opioid development and use. And, then there was a moral panic.

At critical conjunctures along the intersections of history after a catastrophic event, a supposed regeneration takes place that can offer promise and hope for new growth. It happens with wars, diseases, giant storms, and even wildfires. In modern America, after the Storm of 1900 wiped out Galveston, it spawned new growth with the discovery of oil and the building boom in Houston, Texas. In the 1920s, culture and society loosened social norms as women received the vote, smoked on the street, wore flapper outfits and bobbed hair, and attempted to reclaim their bodies. The 1950s likewise marked changes in home ownership along the crabgrass frontier in planned communities like Levittown and consumer culture, like Earl Tupper's ware fashioned in plastic made leftovers a breeze [71]. Thus, the chemistry laboratory in America became that engine of industry and change that had only previously dreamed new products. Hard and soft chemicals took their places as integral to scientific development. Of course, if the Americans were going to beat the Russians at everything from rocketry to washing machines, they would need to harness the latest science. A moral high ground had to be established, and if science would not do it, then reform-minded therapeutic personnel would. Hence what became known as the *goof-ball* panic [72].[34]

---

[34]The term goof-ball was used to describe people's behavior after they took any numbers of types of legal or illegal pharmaceuticals or narcotics (which many considered the same); hence, why moralists perceived this as a wide-spread crisis. The panic was focused on barbiturates, but other pharmaceuticals, either through prescription or over-the-counter, were also targeted. To act like a goof-ball continues to be used in our lexicon today, although the connotation usually describes someone's personality, such as to be goofy or to clown around (as the phrase goes). Goofy was even the same ascribed to a Disney character during the same time period that acted sloppily and spoke with a slow tongue.

Most of moral worrying about the effects of drugs that caused *goofy* behavior after ingesting them, started when the Congress, heavily occupied with chasing McCarthy's Communists and blacklisting everyone from screen writers to directors to stage hands, turned its attention to the drug industry practices of making harmful pharmaceuticals or copying ones that ended up killing patients [72]. Once again, a reactive strategy, which harkened back to Massengill, was applied after thalidomide, a drug given to women during pregnancy to stem *morning sickness*, produced a rash of deformed babies in both Europe and America. Outraged Congressmen, normally occupied with trying to get re-elected, had to act, and they began to finally see the pill mills as manipulators of physicians and their expert medical knowledge. It took time though, and it was not really until the legislative outburst of the Johnson Administration's cajoling in 1965 led to a major alteration in the production of pharmaceuticals [19, 23]. What patients and doctors got was more legislation, called the Drug Abuse Control Amendments that was profitable news to Big Pharma; namely, those prescriptions that were written by educated and trained professionals were not addictive narcotics. That was just the break the firms needed. Now with an expediency to produce and a moral code behind them, they would have free reign to convert morphine-producing laboratories into new divisions geared towards opioid roll-outs. It would be a Pyrrhic victory for the FDA. The laws handed them immense new powers to tract manufacturers and distributors, to bring criminal charges against those who were in possession of drugs without a prescription, and a host of other law enforcement capabilities. But, was the FDA ready for such a move? What if they could not regulate, from a position of authority, the markets where pharmaceuticals were mixed with other drugs that could be found on the street? Additionally, what about those pharmaceutical business channels, which were still ramping up production in pursuit of pain, but also profit? Finally, does not repressing deviance lead to more? Answers were given, but more questions remained, as the jury remained in stasis, and the opioid was unleashed.

The thought probably never crossed Paul Ehrlich's fine chemical mind before his death during the First World War that an era of combinations of synthetic opiates would have such an impact on the post-modern world. Or did it? Though he wrote and spoke about them in the context of fighting diseases, not necessarily synthetic opiates that relieved pain, his prediction was fully embraced by Big Pharma. As they conceived it, with the help along the way from a steady progression that included the work of Dr. Small's Drug Laboratory to the scattered approaches of the federal government in the wake of the *goof-ball* crisis, the 1960s continued to bless the premise from the nineteenth century that if some was good, more would be even better. With an excessive chemical mentality in tow that was coupled with the good intentions of Washington-based law-making, now the firms, like MSD and Purdue Pharma, had all the momentum they would need to gain hegemony. The only chess piece that stood in their way was the laboratory. Humbly standing as a rampart, this was the last place where an opioid roadblock could take place. By the 1980s, the doctor's office and hospitals were spaces where professional healthcare providers could adequately treat or numb patients and professional junkies alike. In response, would the lab possess the necessary mettle to stand strong and repulse a powerful attack that was moving toward the sound of inevitability? As for what the laboratory created, would they merely shrug at the world of pain relief they had made? In a series of conjunctures and with the assistance of politicians, government agencies, marketing specialists, healthcare professionals and last but not least, chemists, the dawn of the opioid empire, and the many combinations and iterations of it, had broken.

References

1. Hatch JL (1903) Cough in pulmonary phthisis. Canadian J Med and Surg 14:lxviii
2. Adams SH (1905–1906) The great fraud. Collier and Son, n.p
3. Kohler R (1982) From medical chemistry to biochemistry: the making of a biomedical discipline. Cambridge University Press, Cambridge, pp 34–39, 121–126
4. Liebenau J (1987) Medical science and medical industry: the formation of the American pharmaceutical industry. Johns Hopkins University Press, Baltimore
5. Warner JH (1991) Ideals of science and their discontents in late-nineteenth century American medicine. Isis 82:454–478
6. Bassi E (2017) An aqueous territory: sailor geographies and New Granada's transimperial greater Caribbean world. Duke University Press, Durham
7. Gabriel J (2014) Medical monopoly: intellectual property rights and the origins of the modern pharmaceutical industry. University of Chicago Press, Chicago, pp 222–227
8. Lanman & Kemp business papers, Hagley Library, Wilmington, Delaware
9. Otero-Cleves A (2017) Foreign machetes and cheap cotton cloth: popular consumers and imported commodities in nineteenth-century Colombia. Hisp Am Hist Rev 97:423–456
10. Courtwright D (1983) The hidden epidemic: opiate addiction and cocaine use in the south, 1860–1920. J South Hist 49:57–72
11. Gortler L (2000) Merck in America: the first 70 years from fine chemicals to pharmaceutical giant. Bull Hist Chem 25:1–9
12. Galambos L, Sturchio J (1994) Transnational investment: the Merck experience, 1891–1925. In: Pohl H (ed) Transnational investment from the 19th century to the present. Franz Steiner Verlag, Stuttgart, Germany
13. Skyhorse Publishing (ed) (2007) Sears catalog, 1897. Skyhorse Publishing, New York, p 26
14. Warner JH (1985) The selective transport of medical knowledge: antebellum American physicians and Parisian medical therapeutics. Bull Hist Med 59:213–231
15. Burrow J (1970) The prescription-drug policies of the American Medical Association in the Progressive Era. In: Blake J (ed) Safeguarding the public: historical aspects of medicinal drug control. Johns Hopkins University Press, Baltimore, p 27
16. Rosenberg C (1979) The therapeutic revolution: medicine, meaning and social change in nineteenth century America. In: Vogel M, Rosenberg C (eds) The therapeutic revolution: essays in the social history of American medicine. University Pennsylvania Press, Philadelphia, pp 3–25
17. Swann J (1988) Academic scientists and the pharmaceutical industry: cooperative research in twentieth-century America. Johns Hopkins University Press, Baltimore
18. Merck & Co. (1908) The Merck report. Rahway, NJ: June
19. Boyle E (2013) Quack medicine: a history of combating health fraud in twentieth-century America. Praeger, Santa Barbara
20. Malleck D (2015) When good drugs go bad: opium, medicine, and the origins of Canada's drug laws. UBC Press, Vancouver
21. Hollon MF (1999) Direct-to-consumer marketing of prescription drugs: creating consumer demand. JAMA 281(4):382–384
22. Acker CJ (1995) From All-Purpose Anodyne to Marker of Deviance: Physicians' Attitudes Toward Opiates from 1890 to 1940. In: Porter R, Teich M (eds) Drugs and Narcotics in History. Cambridge University Press, Cambridge, pp 114–132

23. Starr P (1982) The social transformation of American medicine: the rise of a sovereign profession and the making of a vast industry. Basic Books, New York, p 122
24. Narcotic definition (2017) Merriam-Webster online. https://www.merriam-webster.com/dictionary/narcotic. Accessed 20 Sept 2017
25. Hexamer E (1887) Rosengarten and Sons, manufacturing chemists. Hexamer General Surveys, Volume 22 Plates 2111–2112 https://www.philageohistory.org/rdic-images/view-image.cfm/HGSv22.2111-2112. Accessed 30 Sept 2017
26. Blanc A (1876). Rosengarten & Sons, manufacturing chemists, Philadelphia. [graphic] [Chromolithographs]. https://libwww.freelibrary.org/digital/item/20576. Accessed 30 Sept 2017
27. Burrow J (1977) Organized medicine in the progressive era: the move toward monopoly. Johns Hopkins University Press, Baltimore
28. Geison G (1979) Divided we stand: physiologists and clinicians in the American context. In: Vogel M, Rosenberg C (eds) The therapeutic revolution: essays in the social history of American medicine. University Pennsylvania Press, Philadelphia, pp 67–90
29. Chandler AD (2005) Shaping the industrial century: the remarkable story of the evolution of the modern chemical and pharmaceutical industries. Harvard University Press, Cambridge
30. Lewis H (1947) General problems of laboratory design. J Chem Ed 24:320–323
31. Liebenau J (1985) Scientific ambitions: the pharmaceutical industry, 1900—1920. Pharm in Hist 27:3–11
32. Musto DF (1999) The American disease: origins of narcotic control. Oxford University Press, Oxford
33. Green M (2017) By more than Providence: grand strategy and American power in the Asia Pacific since 1783. Columbia University Press, New York, pp 117–122
34. Liebenau J (1990) Paul Ehrlich as a commercial scientist and research administrator. Med Hist 34:65–78
35. Ferguson N (2006) War of the world: twentieth-century conflict and the descent of the west. Penguin Press, New York
36. Lewenstein B (1989) To improve our knowledge in nature and arts: a history of chemical education in the United States. J Chem Ed. 66:37–44
37. Liebenau J (1984) Industrial R&D in pharmaceutical firms in the early 20th century. Bus Hist 26:329–346
38. Daemmrich A (2004) Pharmacopolitics: drug regulation in the U.S. and Germany. University North Carolina Press, Chapel Hill
39. Marks H (1997) The progress of experiment: science and therapeutic reform in the United States, 1900–1990. Cambridge University Press, Cambridge
40. Rothstein W (1972) American physicians in the nineteenth-century from sects to science. Johns Hopkins University Press, Baltimore
41. Kohler R (1991) Partners in science: foundations and natural scientists, 1900–1945. University Chicago Press, Chicago
42. Rooney S, Campbell JN (2017) How aspirin entered our medicine cabinet. Springer Briefs in the History of Chemistry. Springer, Heidelberg
43. Gross D (2015) Chemical warfare: from the European battlefield to the American laboratory. Distillations Spring. https://www.chemheritage.org/distillations/magazine/chemical-warfare-from-the-european-battlefield-to-the-american-laboratory. Accessed 17 Sept 2017
44. Steen K (2014) The American synthetic organic chemicals industry: war and politics, 1910–1930. University North Carolina Press, Chapel Hill
45. Steen K (1995) Confiscated commerce: American importers of German synthetic organic chemicals, 1914–1929. Hist Tech 12:261–285
46. Steen K (2001) Patents, patriotism, and 'skilled in the art': USA v. the Chemical Foundation Inc., 1923–1926. Isis 92:91–122
47. Noble D (1977) America by design: science, technology, and the rise of corporate capitalism. Alfred A. Knopf, New York
48. Lenoir T (1997) Instituting science: the cultural production of scientific disciplines. Stanford University Press, Stanford, CA

49. Landau R, Achilladelis B, Scriabine A (eds) (1999) Pharmaceutical innovation: revolutionizing human health. Chemical Heritage Press, Philadelphia
50. PBS Website (1998) Opium throughout history. https://www.pbs.org/wgbh/pages/frontline/shows/heroin/etc/history.html. Accessed 10 Sept 2017
51. Lombardo R (2012) Organized crime in Chicago: beyond the mafia. University of Illinois Press, Champaign, IL
52. Ohler N (2017) Blitzed: drugs in the Third Reich. Houghton Mifflin Harcourt, New York, pp 7, 9–10
53. Sullivan N (2016) The prometheus Bomb: the Manhattan project and government in the dark. University of Nebraska Press, Lincoln, NE, pp 69–75
54. Sewall P (2001) Healers in World War II: an oral history of the American Medical Corps. McFarland Publishing, Jefferson, NC
55. Rand A (2005) Atlas Shrugged, Cent edn. Dutton Books, New York
56. Whyte W (2004) The Organizational Man, Rev edn. University of Pennsylvania Press, Philadelphia
57. Angell M (2005) The truth about the drug companies: how they deceive us and what to do about it. Random House, New York
58. Donohue J (2006) A history of drug advertising: the evolving roles of consumers and consumer protection. Milbank Q 84:659–699
59. Puaca LM (2014) Searching for scientific womanpower: technocratic feminism and the politics of national security, 1940–1980. University North Carolina Press, Chapel Hill, p 106
60. Rosen W (2017) Miracle cure: the creation of antibiotics and the birth of modern medicine. Viking Press, New York, pp 86–91
61. White A (1943) Textbook of biochemistry. Third edition (Harrow, Benjamin) J Chem Ed 20:520
62. Mosettig E (1959) Lyndon Frederick Small, 1897–1957: a biographical memoir. Nat Acad Sci, Washington DC
63. Acker CJ (2003) Take as directed: the dilemmas of regulating addictive analgesics and other psychoactive drugs. In: Meldrum M (ed) Opioids and pain relief: a historical perspective. IASP Press, Washington DC, pp 35–55
64. Wimmer W (1998) Innovation in the German pharmaceutical industry, 1880 to 1920. In: Homburg E et al (eds) The chemical industry in Europe, 1850–1914: industrial growth, pollution, and professionalization. Kluwer Academic Publishers, Dordrecht, pp 281–292
65. Acker CJ (1995) Addiction and the laboratory: the work of the National Research Council's Committee on drug addiction, 1928–1939. Isis 86:167–193
66. Acker CJ (2001) Creating the American junkie: addiction research in the classic era of narcotic control. Johns Hopkins University Press, Baltimore
67. Scientific News and Notes (1929) Science 70:401–404
68. National Research Council (U.S.) (1941) Report of committee on drug addiction, 1929–1941. Committee on drug addiction, National Research Council, Washington DC
69. Soares JX, Alves EA, Silva AMN, de Figueiredo NG, Neves JF, Cravo SM, Rangel M, Netto ADP, Carvalho F, Dinis-Oliveira RJ, Afonso CM (2017) Street-like synthesis of Krokodil results in the formation of an enlarged cluster of known and new Morphinans. Chem Res Tox 30:1609–1621
70. Bowden M (2003) Pharmaceutical achievers: the human face of pharmaceutical research. Chemical Heritage Press, Philadelphia
71. Cohen L (2003) A consumers' republic: the politics of mass consumption in postwar America. Vintage Books, New York
72. Rasmussen N (2012) Goofball panic: barbiturates, "dangerous" and addictive drugs, and the regulation of medicine in postwar America. In: Greene J, Watkins E (eds) Prescribed: writing, filling, using, and abusing the prescription in modern America. Johns Hopkins University Press, Baltimore, MD, pp 23–45
73. Anon (1952) Photograph and biography of Lyndon Frederick Small. J Org Chem 17:i–ii

Chapter 4
Epilogue: Opioid Heterotopias

4.1 Sackler's Ghost: The Tale of OxyContin's Failed Chemistry

> The country which is in advance of the rest of the world in chemistry will also be foremost in wealth and in general prosperity.
>
> —William Ramsay, c. 1900 [1]

There was once a large hegemonic empire whose people possessed a great history. They had waged wars of conquest and defeated many enemies that threatened their way of life. Despite this, they verged on the declaration of a state of emergency. Over the past decade or so it was perceived that the addiction abuse rates of a certain drug had grown exponentially after a previous period of supposed tranquility when laws were passed banning the substance. Everyone in positions of power remained confident in their approach to the situation. However, debate ensued over why health experts had sanctioned the use of the drug. In response, the state's government had claimed that it has always known the threat existed; but, unbeknownst to the people, a robust trading business that fueled the economy was always deemed more important. Its leadership gave blusterous speeches about protecting those that were vulnerable. The emperor of the land issued a rash of edicts, and even appointed a *czar* to handle the situation. The entity responsible for the flooding of the market in itself possessed a powerful lobby, and its tentacles reached into the pockets of consumers and politicians alike. One could say that deregulation of trade had made oversight difficult to maintain. Some agents of the government that were tasked with policing districts looking for abuses by merchants were hamstrung by an inefficient bureaucracy and a faraway capital city that legislated based on favor. Mercantile and political stakes were high as the crisis escalated to an all-out war; the only possibility

J. N. Campbell and S. M. Rooney, *A Time-Release History of the Opioid Epidemic*, SpringerBriefs in History of Chemistry, https://doi.org/10.1007/978-3-319-91788-7_4

was the drafting of some kind of accord in order to bring the situation to a cessation. By then, it was too late [2].[1]

The story related here is not about the United States per se, but it could be historically applicable. It actually captures events from the mid-nineteenth century in Qing China and the international manipulations utilized by the British. The former was grappling with what would become known as the Opium War, while the latter wanted access to markets for its India tea in exchange for Chinese export porcelain. Lessons from the past, whether they are learned or unlearned, provide valuable contexts in which to understand heterotopias, such as the chemistry laboratory. Tracing the long history of opioids in the United States offers perspective, especially when considering the events in China and linking it to the current crisis. Once again, the role of the laboratory in designing an alkaloid proved how it could affect society. In the long run, it would serve as a cautionary chemical tale. The story behind Purdue Pharma's OxyContin, a time-release drug that was part of the movement to end pain in the 1990s, actually had a much longer and involved history.

In the wake of the legislative initiatives by the federal government in the late 1960s and 1970s, two movements that were drug-related began to unfold. The first, previously discussed, entailed the clear separation between pharmaceuticals and narcotics; while the second involved the expansion of street drugs, especially in the form of heroin. Dreser's revenge, perhaps? We have seen this before the First World War and after the Second. When certain drugs are banned, others rise. Measuring use still remains an issue [3].[2] Whether spawned by small-time crooks, wannabe mafia-types (real or otherwise), or those seeking fortunes in economic uncertain times, the pharmaceutical world was essentially handed all of the tools it would need to cook up a disaster. These were short-term developments. In the long-term, chemically-speaking, what was even more significant was the heritage established by the laboratory and how the management of Big Pharma companies would interpret new opportunities to court customers. All of these elements, including cultural standards that had long been in place within healthcare delivery geography, would converge into the making of another opiate/opioid-related crisis [4].

It all began during the fire of war. Yet, it was not in Iraq, Rwanda, Kosovo, Somalia or any other country brimming with conflict during the 1990s; rather, it was during the destruction of the 1910s, the First World War. To no surprise, based on their chemical heterotopian history, in 1916, the same year as the Somme where so many

---

[1]The journalist and social commentator G. K. Chesterton argued in *The Medical Mistake, What's Wrong with the World* (London, 1910) that healthcare delivery should only be used for restoring the normal human body. Of course, the invention of public health and its mission to eradicate the individual case in the name of social progress, which began during the Progressive Era and elevated to new heights during the New Deal, continues to be debated as the opioid crisis unfolds. How the state reacts to issues such as these begs the continued question of, can medical problems ever be truly solved by altering the social order or controlling the chemistry? This seems wholly incompatible with one in which people possess moral relationships.

[2]It is the opinion of these authors that the topics of measurement of use might be an interesting line of questioning. For instance, is it possible that since numbers of those addicted today looks inflated simply because statistics simply do not exist for past periods in the history of the United States?

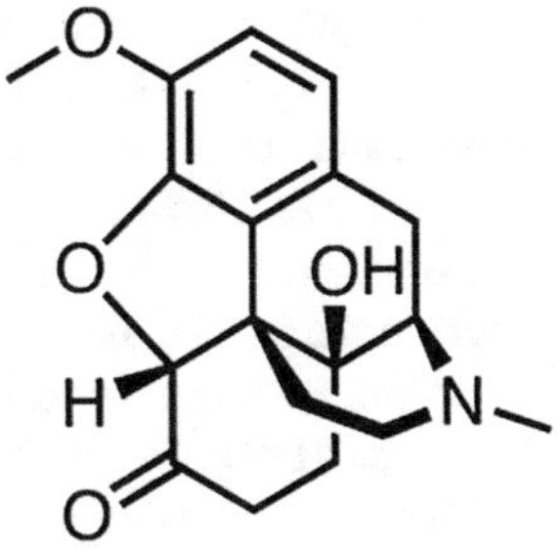

Fig. 4.1 Oxycodone (1917)

perished, two German chemists worked with thebaine. Named for the ancient city of Thebes in Egypt, where writings concerning opium-use thousands of years ago were found, the base that the chemists worked with showed promise since it was closely connected to morphine. Later, it would be used in Small's development of Desomorphine(the precursor to the Russian Krokodil) in the United States. In the meantime, what they synthesized and christened Oxycodone, would reverberate throughout the century to this day. While chemically similar to morphine and codeine, thebaine strikes a major difference in that it stimulates instead of depressing the body. The German creators, namely Martin Freund and Edmund Speyer were employed at the biomedical juggernaut at the University of Frankfurt and inspired by Paul Ehrlich and his vision for a chemical future, achieved success.[3] They took the naturally occurring synthetic, thebaine, not necessarily setting out to produce a pharmacological or therapeutic miracle, and got one. By introducing a 14-hydroxyl group, the advanced laboratory struck gold when it simultaneously boosted the potency of the morphine and codeine analogues (Fig. 4.1). Carl Mannich and Hellene Lowenheim, working for the German pharmaceutical company Knoll, synthesized the semisynthetic narcotic from codeine that later became known more popularly as Hydrocodone. Today, it is more efficiently produced from thebaine and mixed with acetaminophen. This medicinal mixture is commonly known as Vicodin, a drug that has garnered both the ire and applause of patient and consumer alike. Knoll, as so many firms before had done along the drug timeline, held the chemical scalability card and though war was devastating the Second Reich, production soared [5].

The interwar years saw Oxycodone undergo several incarnations. One, in particular, that hit the European markets, the same year Small came to the University of Virginia to start his drug addiction laboratory, was at this time that major pharmaceutical company that had started morphine production exactly 100 years before in Darmstadt, Germany—E. Merck. Relying on a chemical heritage they combined scopolamine, ephedrine, and oxycodone under the German initials, SEE, to form a highly-addictive pharmaceutical. The clever firm took a letter from each of the main compounds present; thus, bringing Ehrlich's prophecy that combined drugs

---

[3]Martin Freund (1863–1920) and Edmund Speyer (1878–1942) developed a close working relationship as they worked with Thebaine. Freund was eulogized by Speyer after his death in 1920. Once the Nazis rose to power in the 1930s, Speyer was targeted and banned from teaching because of his Jewish faith. He died in a ghetto in 1942 due to a "weakness of heart."

worked better in concert into formal practice.[4] After Oxycodone officially arrived in America in 1939, it lay dormant for some time; perhaps the competition, other drug options, and the influence of the street drove it underground as the federal government classified it as a Schedule II drug [6].[5]

From there a supposed villain, a person so full of avarice and a preference for chicanery, emerged. He had both a medical background and a penchant for understanding people's wants and desires through his training in psychology and neuroscience. Arthur Sackler's stock as both a progenitor of pharmaceuticals and as an exploiter of human beings today has hit rock bottom. His family, including two brothers who were also trained in medicine, purchased in 1952 the Purdue Frederick Company that was started in the late nineteenth century by doctors John Purdue Gray and George Frederick Bingham in New York City. They were mainly known as collectors of fine art and from signage on the walls of the major museums they had generously gifted those treasures. Sackler was unique to be sure, but not because he combined science, advertising, and an understanding of market strategies [7].[6] Rather, he was the recipient of a time-release moment in healthcare delivery and drug design. Advertising, in the era of *Mad Men*, was being revolutionized into a science of its own, and with a cornucopia of new miracle drugs that were rolled out, it seemed each day, the recipe was brewing for taking those branded items directly to the consumers.[7] Of course, selling what the laboratory created was nothing new, after all we have already seen that the trade journals marketed heroin at the turn-of-the-century through direct and indirect means.[8] In the postwar era, with television, print culture, and access at an all-time high, the stage was set for product placement and a complete isolation of those all-important chemists that developed the latest pharmaceuticals in their chemical

---

[4]Ephedrine in its natural form has a long history. For instance, in traditional Chinese medicine, má huáng has been used as a treatment for asthma and bronchitis for centuries.

[5]A Schedule II translated to a category of drugs considered to have a strong potential for abuse or addiction, but that have legitimate medical use. Among the substances so classified by the Drug Enforcement Agency are morphine, cocaine, pentobarbital, oxycodone, and methadone. Yet, now, even the present crisis brings into question what *legitimate medical use* will mean into the future.

[6]Arthur Sackler (1913–1987) had a long career that included along with his brothers, the publishing of over 140 research papers particularly pertaining to the mental illness. He took a job at the William Douglas MacAdams Advertising Agency in order to finance his medical education. Eventually, he worked his way up to become the principal owner. In 1958, he founded the Laboratories for Therapeutic Research, a nonprofit basic research center at the Brooklyn College of Pharmacy of Long Island University, and served as its director until 1983. With an art history background, he travelled the world acquiring antiques (particularly Chinese Art) that would become part of major collections at museums including the Met in New York City and the Smithsonian Institution in Washington D.C.

[7]Matthew Weiner's AMC Series *Mad Men* (2007–2015) portrayed the raucous world of advertising in the 1950s and 1960s. See, http://www.newyorker.com/culture/cultural-comment/the-original-resonant-existentially-brilliant-mad-men-finale (accessed October 21, 2017).

[8]Several authors displaying credentials of high journalistic integrity have bitten into this notion that Sackler was a revolutionary advertiser. He was, but it should be noted that there were historical examples of medical journals touting products in professional journals fifty years before he came up with the idea. Perhaps Patrick Keefe of the *New Yorker* and Christopher Glazek of *Esquire* only perceive the recent past as the most relevant [11, 12].

heterotopias. More than ever, the idea of white lab-clad scientists ensconced in seclusion was becoming an image ingrained in popular imagination as patients morphed into consumers even further [8, 9].

The Sackler Family's investment in Purdue Pharma, as it became known, was rather unspectacular for over three decades. They watched as other more powerful firms like MSD, Eli Lilly, and Parke-Davis spun new drug after new drug. The one major victory Purdue had in its chemical arsenal at this point was a product called MS Contin (which ironically was short for continuous) that was purchased from a British company the firm had acquired. However, by the late 1980s the patent was soon to expire, and the company started looking for a replacement. The choice to replace it was an opioid drug that was released in 1995 and became known as OxyContin. Prior to this, in 1972, several companies were working on the concept of time-release formulations. Science fiction writers had long mused over the future of drugs, where a small pill at different intervals would release healing medications into the bloodstream. Now chemists working in those scaled laboratories conceived that an older compound, one that was relegated to the back shelves should make its triumphant return. Taking Oxycodone, the same one that those German chemists had stumbled upon in 1916, and reformatting it, Purdue developed OxyContin as a non-addictive opioid substitute for heroin, yet it acted chemically in a similar fashion [10]. Oxycodone and its long heritage was just what the Sacklers needed; an inexpensive to produce pharmaceutical, and one that later would be used in other drugs like Percodan (oxycodone and aspirin), released in 1950, and Percocet (oxycodone and Tylenol), released in 1974.

All seemed right in the world. Doctors, eager to prescribe a miracle for those suffering from chronic pain wanted relief, as changing standards about suffering and healthcare delivery continued to support the notion that total mastery was within their grasp. Likewise, firms swimming against the competitive stream of capitalism wanted to separate themselves from other products. They would achieve this by any means necessary as they began to work with the next generation of companies that followed in the footsteps of Lanman & Kemp in the drug distributing business, like McKesson, Cardinal Health, and AmerisourceBergen. These powerful links to clinicians would have dire consequences as the dawn of the new century unfolded. Medical care was being revolutionized with government-sponsorship in tow; patient-advocacy groups were convinced that risks for addiction were low. A perfect storm developed in the doctor's office, but on the street, as with heroin almost a century before, painkillers were found to be readily available. Somehow, and we are not sure exactly where, someone discovered that if you crushed an OxyContin, the chemical stability, so delicately conceived within the laboratory would be broken down and a speedy high would be offered to a ready user. Soon, the nursing homes and hospital rooms had patients turn into veritable drug dealers who sought to pawn their pills to those on the street. Chronic pain was thus tolerated in the name of supporting a family. Once again, the street heterotopia looked more innovative than the Big Pharma one [13].

What makes OxyContin different from other opioid pain killers is its supposed time-release formula. Since the powerful drug had a patented delayed-absorption mechanism, it was thought to be safer than competitive painkillers. As a matter of fact, Purdue Pharma never conducted clinical studies on how addictive the drug might be. It appeared that the drug company convinced the FDA to approve the drug without these studies claiming the time-release formula made the drug less prone to addiction. Coincidentally, Dr. Curtis Wright, the FDA examiner who oversaw the approval process of OxyContin, left the agency soon after and took a job at none other than, Purdue Pharma. Through aggressive advertising and misinformation about the potential for abuse of OxyContin upon its release in 1995, it was hailed as a medical breakthrough, a long-lasting narcotic that could help patients suffering from moderate to severe pain. The drug became a blockbuster, and has reportedly generated some thirty-five billion dollars in revenue for Purdue.[9] Yet, what also needs to be acknowledged is that once OxyContin was broken down, like Krokodil in Russia, it was incorporated with other drugs (like Methadone or heroin), whatever was on hand, into a bastardization of Ehrlich's combination drug theory, which has become a recipe for destruction [14].[10]

Why is it that the complex developments, the well-constructed theories, and the most carefully planned explanations fall well-short when placed alongside cultural considerations? How is that the chemists, behind such a complex time-release pharmaceutical with such a long history, could have their drug taken, ground into a powder, and its intricate design be manipulated by street gangs and those that sought to do harm to others or even themselves—a failure of chemistry? Perhaps part of the answer lies in the nature of culture itself. That is, the chemical heterotopia, with its long tradition and history is not one space comprised of white lab coats and pristine clear glass. The laboratory was and still is an iteration that comes in many forms and guises. It can be found at universities, in large chemical companies, on the street and

---

[9] At the time of the publication of this volume, Purdue Pharma has been sued and severely penalized for its practices surrounding the development and marketing of Oxycontin, which includes jail time for some of the executives involved in the sale of the drug. For instance, as recently as January 2017 the city of Everett, Washington sued Purdue based on increased costs for the city from the use of Oxycontin. They also contended that the firm did not intervene when they noted abnormal patterns of sale of their product, per the agreement in the 2007. The allegations include not following legal agreements to track suspicious excess ordering or potential illegal usage. False clinics were uncovered by unscrupulous doctors using homeless individuals as 'patients' to purchase Oxycontin, who then in turn would sell them to the citizens of Everett was the factual basis of the suit. The illegal sale of the drug out of legal pharmacies based in Los Angeles with distributions points in Everett is also part of the larger story. Purdue did not contact the DEA for years despite knowing of the practice and the overuse and sale of their product. The suit is asking for a yet to be determined reimbursement related to costs of policing, housing, health care, rehabilitation, criminal justice system, park and recreations department, as well as to the loss of life or compromised quality of life of the citizens of the city directly.

[10] In 2010, in response to lawsuits and bad press, Purdue's laboratories rolled out a new form of OxyContin called OxyNeo. This drug was not crushable and even when liquid was added it would turn into an insoluble glob. Already abusers have begun to turn to other combinations of drugs in response. Other companies like Endo and Janssen have developed their own versions of abuse-deterrent products. Time will tell how this impacts the current opioid crisis. Ironically, the latter was founded by Paul Janssen whose company was the first to synthesize fentanyl in 1959, which was used as a general anesthetic under the trade name Sublimaze in the 1960s. Now, it has become one of the most powerful opioids in America. For more discussion of this topic, see Foreman J (2014) *A Nation in Pain: healing Our Biggest Health Problem*. Oxford University Press, Oxford, p. 172–174.

in the back alleyways, and even in the kitchens and in double-wides. The one constant found in all of these are the human beings that construct the chemistry, deem what is usable and what is not, and are the ones that can make truly horrendous choices or apply the very highest standards to their craft. The pristine lab, the professional one, is just as capable, as seen in the case of Purdue Pharma, of behaving adroitly in some cases or conversely in others; thus, effecting deception and intrigue in its midst. It is not necessarily the ghosts of the Arthur Sacklers that are the ones that haunt us; rather, the chemists' experiments and what they uncover that reverberates throughout history and that matter most. When combined with different cultural implications, understanding the chemical past can inform us just as well in the present. What makes the laboratory such an important crossroads is what it has always been; that being, a place where human beings who are flawed and brilliant, hopeful and shameful, avaricious and noble, engage in a process of creation that imposes a morality to things that are supposedly scientific.

4.2   Time-Release: The Story Continues

Time is the best appraiser of scientific work, and I am aware that an industrial discovery rarely produces all its fruit in the hands of its first inventor.
—Louis Pasteur, c 1860 [15]

In general chemical terms, if they exist, an agonist is a compound that binds to a receptor and activates it in order to produce a biological response. Metaphorically, the historical justification for the recent rise of opioids, and the subsequent deaths they are blamed for, is quite similar. Scholars and journalists seem quite taken with accepting the binding fact that before the 1970s, doctors' desire to not create a rash of addicts was the overriding explanation for why opioids were kept at bay. Their proclivity for sleuthing has led them to believe that one of the major smoking guns that changed everything has to be a hundred-word letter to the editor published in a short piece penned by a doctor and his assistant in a 1980 issue of the *New England Journal of Medicine* (*NEJM*), which related a low-addiction rate in a recent study they conducted (Fig. 4.2).

Thirty-seven years later in another *NEJM* issue, a consortium of doctors performed what they dubbed as a bibliometric analysis of the citations that followed the recommendation that opioid addictions were low [17]. What resulted in the publishing of this study was an interesting instance once again of that old game of telephone. News agencies of every sort from *National Public Radio* to the *Daily Beast* to the *Washington Post* burned up the fiber sending out assessments and offering scathing editorials. Finally, here was the root; here was where everything medically went awry. Most journalists burned in effigy the doctor, Hershel Jick, and took successive healthcare professionals to task for their dewy-eyed idealism and slipshod prescribing. The arc of how the situation developed in the wake of the 1980 Letter and the reasons behind why these news agencies were so quick to draw conclusions after the 2017 study was released, presents a fascinating example.

Especially, when juxtaposed with the repeated reprinting of the heroin testimonial of Dr. John Leffingwell Hatch from the turn of the twentieth century. In many ways, the same process of how medical knowledge was disseminated applied, even though it appeared that the 2017 study was based in the latest online media gathering mechanisms [18]. What this situation tells us is that once again the short-term history can drive the desire for answers in the midst of a crisis, much more so than the analogy of the slow-moving train that steadily gains speed over time. The public and the professional communities look for low-hanging fruit, and the media is all too willing to oblige with a rickety-old stepstool.

The story of opioids, when a wide-angle historical lens is applied to this imperfect resemblance, reflects the integral role played by a diverse set of Foucault-inspired chemical heterotopias. What can be surmised, over the long history from alkaloid to opiate to synthetic opioid, is that like a time-release pill, they are directly impacted first and foremost by the machinations and inner-workings of the chemistry laboratory [19]. Actors related to the history of opioids were disparate, but united in a past linked across space and time. From the Lavoisiers to Sertürner; from Liebig to Pinkham; from Lanman to George Merck; or from Ehrlich to Small, each *chemist* occupied a place on the laboratory stage in some form or fashion. Some were creators, others were gatekeepers, and a few, even adopted a social mission for their work, while the rest were unsuspecting pawns or avaricious profiteers. Over the generations, healthcare delivery professionals dating back to the age of nostrums and cooks in the nineteenth century have continued to defend their position as the avatars of the patients and their pain. Yet, we have also seen the ascendancy of a powerful pharmaceutical industry hailing from the lands of coal tar and chemistry, which sought to mold a populace into good consumers. The line between them has blurred, isolating the laboratory, muddying the polished floors of hospitals, and the sanctity

ADDICTION RARE IN PATIENTS TREATED WITH NARCOTICS

To the Editor: Recently, we examined our current files to determine the incidence of narcotic addiction in 39,946 hospitalized medical patients' who were monitored consecutively. Although there were 11,882 patients who received at least one narcotic preparation, there were only four cases of reasonably well documented addiction in patients who had a history of addiction. The addiction was considered major in only one instance. The drugs implicated were meperidine in two patients, Percodan in one, and hydromorphone in one. We conclude that despite widespread use of narcotic drugs in hospitals, the development of addiction is rare in medical patients with no history of addiction.

JANE PORTER
HERSHEL JICK, M.D.
Boston Collaborative Drug
Surveillance Program
Boston University Medical Center

Waltham, MA 02154

Fig. 4.2 Facsimile of 1980 Letter, *New England Journal of Medicine* [16]

of the clinics. Moving forward, with an opioid death count continuing to inflate, a host of new worlds opening where organs are more plentiful and cops deploy sprays to induce life again, hope hangs by a thread as to whether the United States can control the policies that are meant to protect its own chemical future [20]. One piece of commentary that can be offered is to think of the authority established by the laboratory for over 200 years. Simply put, more can be done. Beginning with plant chemistry, some of the most significant members of the pharmaceutical community, can harness compounds and produce technological advancements like synthetic time-release drugs. Until those in positions of influence grasp the fact that the science behind the development of opioids endeavors to serve the people; the chemistry will *continue* to be marginalized in favor of marketing and profit, all at the expense of humanity. Only the next historical time-release rendition from any of the chemical heterotopias will slow this inexorable runaway train or render the next version of a dystopian dream.

References

1. Travers M (1956) The life of Sir William Ramsay. Arnold, London
2. Pomeranz K (2001) The great divergence: China, Europe, and the making of the modern world economy. Princeton Univ Press, Princeton, NJ
3. Acker CJ (2001) Creating the American Junkie: addiction research in the classic era of narcotic control. Johns Hopkins Univ Press, Baltimore, pp 135–139
4. Moynihan R, Cassels A (2006) Selling sickness: how the world's biggest pharmaceutical companies are turning us all into patients. Nation Books, New York, pp 196–200
5. Daemmrich A (2005) Pharmacopolitics: drug regulation in the United States and Germany. University of North Carolina Press, Chapel Hill, pp 6, 9
6. Beauchamp T, Childress J (2012) Principles of biomedical ethics, 7th ed. Oxford University Press, Oxford, pp 94, 284
7. Glueck G (1987) Dr. Arthur Sackler dies at 73; Philanthropist and art patron. NYT May 27 http://www.nytimes.com/1987/05/27/obituaries/dr-arthur-sackler-dies-at-73-philanthropist-and-art-patron.html. Accessed 25 September 2017
8. Rosen W (2017) Miracle cure: the creation of antibiotics and the birth of modern medicine. Viking Press, New York, pp 227–228
9. Rooney S, Campbell JN (2017) How aspirin entered our medicine cabinet. Springer, Heidelberg, Springer Briefs in the History of Chemistry, p 42
10. Greene JA, Watkins ES (eds) (2012) Prescribed: writing, filling, using, and abusing the prescription in Modern America. Johns Hopkins University Press, Baltimore, MD
11. Keefe P (2017) The family that built an empire of pain. New Yorker Oct 30 https://www.newyorker.com/magazine/2017/10/30/the-family-that-built-an-empire-of-pain Accessed 31 Oct 2017
12. Glazek C (2017) The secretive family making billions from the Opioid Crisis. Oct 16 www.esquire.com/news-politics/a12775932/sackler-family-oxycontin/. Accessed 31 Oct 2017
13. Acker CJ (2003) Take as directed: The dilemmas of regulating addictive analgesics and other psychoactive drugs. In: Meldrum M (ed) Opioids and pain relief: a historical perspective. IASP Press, Washington DC, pp 35–55
14. Goldacre B (2014) Bad pharma: how drug companies mislead doctors and harm patients. Farrar, Straus and Giroux, New York, pp 174, 313
15. Robbins L (2001) Louis Pasteur: and the hidden world of microbes. Oxford University Press, Oxford

16. Porter J, Jick H (1980) Addiction rare in patients treated with narcotics. N Engl J Med 302:123
17. Leung P, Macdonald EM, Stanbrook MB, Dhalla IA, Juurlink DN (2017) A 1980 letter on the risk of opioid addiction. N Engl J Med 376:2194–2195
18. Burch D (2010) Taking the medicine: a short history of medicine's beautiful idea, and our difficulty swallowing it. Random House UK, London, p 47
19. Abramson J (2008) Overdosed America: the broken promise of American medicine. Harper Perennial, New York, pp 102–103
20. Werth B (2014) The antidote: inside the world of new pharma. Simon & Schuster, New York, p 3

Index

A

Adams, Samuel Hopkins, *Collier's*, 59, 78
Advertisements
  heroin, 49, 51, 53–55, 67, 70–73, 79, 82–85
  opiates, 1, 5, 34, 36, 38, 61, 63, 65, 68, 73,
    75, 77, 80, 83, 86, 92, 94, 102, 108,
    112, 114, 117
  opioids, 5–7, 29, 49, 70, 75, 89, 96, 99,
    103, 104, 114, 126
  patent medicines, 31–34, 51, 78, 86
Alien Property Custodian (APC)
  Merck sale, 90, 92
Alkaloids
  chemical make-up of, 6, 12, 23, 37, 70, 71,
    73, 102
  history of, 18, 65
American Medical Association (AMA)
  Journal of, *see* medical journals
American Pharmaceutical Association (APhA),
    29, 32, 33, 36
Apothecary
  assistant to, 11, 20
  history of, 13
Aspirin
  history of, 48, 51, 55
  making of, 66

B

Bache, Alexander
  Lazzaroni, 27
Bayer, Farbenfabriken
  acetylsalicyclic acid (ASA), 50
  salicyclic acid, 50
Bayer, Friedrich, 47–49, 51, 53, 54, 65–68, 73,
    78, 84, 93

Bernard, Claude, 75, 76
Big Pharma
  history of, 7, 61, 68, 82
Biochemistry
  creation of university programs in, 42, 83,
    86, 94, 105
Bureau of Chemistry (BOC)
  founding of, 86
Burroughs Wellcome & Company
  use of the term *tabloid*, 2

C

Chandler Chemistry Building
  Lehigh University, 38
Chase, Dr. Alvin Wood
  methods, 35
Coal tar, *see* pharmaceuticals
Codeine
  history of, 113
Conant, James B.
  MIT, 108
Congress of Vienna, 19

D

Darmstadt, Germany, 23–25, 66, 90, 123
David, Jacque Louis
  painting of Lavoisiers, 10, 11
Doctors, professionalization, *see* hospitals;
    laboratories
Dreser, Heinrich
  making of heroin, 51, 55
Drug distributors, *see* pharmaceuticals
Duisberg, Carl
  history at Bayer, 48, 49, 66, 82
  IG Farben and Carl Bosch, 88, 102

© The Author(s) 2018
J. N. Campbell and S. M. Rooney, *A Time-Release History of the Opioid Epidemic*,
SpringerBriefs in History of Chemistry, https://doi.org/10.1007/978-3-319-91788-7

Dye companies
    German, 48, 60, 62

E
Ehrlich, Paul
    chemical theories and practices, 84, 86, 126
    institute, 84, 85
    magic bullet, 112
    Nobel Prize, 84
Eichengrün, Arthur, 49, 50
E. Merck
    history of, 23, 25, 27, 46, 48, 55

F
Federal Trade Commission (FTC)
    jurisdiction of, 96
First World War
    history of, 6, 82
    influence on chemical industry in Europe
        and America, 55
    World War I, 87
Food, Drug and Cosmetic Act of 1938
    creation of, 92, 96

G
Galt, John
    Who is?, 100
Garvin, Francis
    APC, 90
Gay-Lussac, Joseph
    alkaloids, 18, 19, 25, 55
    Gay-Lussac Law, 18
    work with Pierre-Jean Robiquet, 19
Giebert, Georg Christian, 44, 45
Glyco-Heroin
    advertisements for, 67, 72

H
Harrison Narcotics Act of 1914, 74, 82
Hatch, Dr. J. Leffingwell, 59, 72, 73, 75, 76,
    128
Heroin Act of 1924, 93
Heroin, Bayer
    origin and history of, 49, 52, 53, 65, 66, 83,
        85
Heterotopia
    definition of, 3, 4
    Michael Foucault's theoretical framework,
        3
Himmelsbach, Clifton
    development of scale for withdrawal
        symptoms, 116
Hofmann, August Wilhelm (von), 42, 46, 47,
    49, 55

Hospitals
    examples of, 61, 63, 83, 110, 112, 115, 117
    laboratories, 86, 87, 128
    Lexington, 110, 112, 114–116
    professionalization of, 6, 32, 61, 77
Hunter, Dr. Charles
    hypodermic needle debate, 28
Hunt, Reid
    Committee on Drug Addiction role, 111

I
IG Farben
    Nazis use of, 103
International Opium Commission
    founding of, 81

K
Kebler, Lyman F., 79
Krokodil
    connection to earlier synthetic opioids, 113,
        123, 126

L
Laboratories
    different types of chemistry, 4–7, 9, 10, 14,
        23, 25–27, 30, 32, 33, 35–37, 39, 40,
        42, 45, 47–49, 51, 53, 55, 60, 61, 63,
        66–68, 70, 71, 73, 76, 77, 79, 81, 84,
        86, 87, 91, 94, 96, 100, 107, 109, 111,
        116, 117, 126, 128
Lanman, David
    aqueous connections, 63
    as distributor of chemicals, 62, 63
    Florida Water, 62
    Lanman & Kemp partnership, 37, 62–65,
        67, 68, 125
Laudanum
    history of, 6, 14, 29, 34–36
    making of, 36, 51
Lavoisier, Antoine
    death of, 9, 13
    new laboratory, 9, 10, 12, 25
Lavoisier, Marie-Anne
    defense of husband, 9
    role in laboratory, 10
Liebig, Justus von
    chemical philosophy, 37
    creation of beef extract, 39, 45, 53
    dream of, 46, 48, 55, 79, 83
    laboratory at the University of Giessen
        (images), 40–42
    students, 42–44, 76, 105

M

Massengill deaths
    findings, 94–96, 99, 117
    Harold Watkins, 95
Materia medica
    definition of, 12, 60, 62, 65, 84
Media
    coverage by, 128
    opioid crisis, 7
Medical journals
    examples of, 13, 17, 20, 53, 61, 66, 67, 69,
        70, 72, 76, 107, 110, 116, 127
    history of, 60, 71, 73
Meissner, Carl Friedrich Wilhelm
    first use of the term Alkaloid, 20, 21
Merck
    in America, 27, 60, 65, 67
    in Germany, 23, 24
    sale of alkaloids and other pharmaceuticals,
        23, 24, 55
Merck & Co. (MSD), 66, 90–93
Merck, George
    name change, 66, 89, 90, 128
Merck, Heinrich Emanuel
    Engel-Apotheke, 23
Merck's *Index*, 71, 78
Methadone
    history of, 7, 102, 103, 126
    making of, 83, 102
Morphine
    history and invention of, 6, 7, 13, 16–20,
        25, 27, 28, 30, 38, 39, 51–54, 60–62,
        66, 73, 79, 87, 92, 93, 98, 99, 103, 113,
        117, 123
    soldier's disease myth, 28
Mueller, JoAnne
    home laboratory, 107

N

Narcotics
    Division of Narcotics, 94
    history of, 13, 80, 81, 86
    making of, 74
    U.S. Treasury Division of Narcotics, 93
Neurotransmitters
    effects, 25
    examples of, 22
New England Journal of Medicine (*NEJM*),
        1980 Letter, Dr. Hershel Jick, Jane
        Porter, 127, 128
Nostrums, *see* patent medicines

O

Opiates
    Civil War use of, *see* patent medicines
    origins and making of, 5, 20, 29, 33, 36, 70,
        78, 81, 99, 101, 104
Opioids
    chemistry laboratories that make, 100, 101
    death rates, 5
    origin of the word, 1, 2, 5, 7, 27
    semi-synthetics, 93
    synthetics, 102, 103
    time-release, 7
    uses of, 6, 96
Opium
    laws, 81, 113
    origin and history of, 6
    usage, 6, 13, 14, 32
Opium Wars
    China, 6, 13, 25, 81
    Great Britain, 32
    history of, 122
    Lin Zexu, 32
Organic chemistry
    origins and development of, 12, 20, 25, 42,
        49, 55, 60, 70, 75, 79, 86, 105
Oxycodone
    Carl Mannich and Hellene Lowenheim, 123
    history of, 83, 123–125
OxyContin
    Arthur Sackler, 124, 127
    making and marketing of, 7, 122, 125, 126

P

Palmer, A. Mitchell
    APC, 89, 90
    Palmer Raids, 89, 90
Parke-Davis
    making of pharmaceuticals, 68, 100, 125
Patent medicines
    definition of, 6
    history of, 26
    marketing of, 29
    use of, 34
Patents
    history of, 29, 32, 33, 36, 68, 69, 79, 86, 89,
        90, 98, 103, 113, 125
    laws pertaining to, 54
Pharmaceuticals
    alkaloids, 23
    AmerisourceBergen, 60, 125
    Cardinal Health, 60, 125

companies, 5, 17, 29, 45, 68, 78, 85, 100,
      101, 105
  distributors of, 6, 17, 38
  examples of distributors, 29, 38, 44, 60, 65,
      69
  history of, 5
  marketing of, 93
  McKesson, *see* patents
  sales of, 61
Pinkham, Lydia
  advertisements, 30
  letter writing, 31
  patent medicine trade, 29
  Vegetable Compound, 29, 30, 32
Plant chemistry
  connection to biochemistry, 24
  origins of, 10, 12–14, 20, 129
Purdue Pharma
  making of OxyContin, 122, 125–127
Pure Food and Drug Act 1906, 77, 78, 86

R
Radam, William
  Microbe Killer, 26
Rand, Ayn
  Atlas Shrugged, 100
Rosengarten & Sons
  making of chemicals, 37, 74, 75, 91

S
Sackler, Arthur
  art world of, 124
  OxyContin, 125
  Purdue Pharma, 125
  the making of and marketing of plans of,
      126

Saint James Society
  samples of heroin through the mail, 93
Science and Technology Studies (STS)
  types of scholarship, 5
Sears and Roebuck Co.
  Catalog, 68
  selling of morphine, 68
Sertürner, Freidrich Wilhelm
  alkaloids, 12, 24, 25, 42
  Annalen der Physik, 19
  dog subject, 16, 25, 54, 103
  morphine, 16, 20, 25, 102, 103
Small, Lyndon
  cigar boxes, 112
  Desomorphine, 112, 113, 123
  Drug Addiction Laboratory, 108, 111, 123
  linked to krokodil, 113, 123, 126
  University of Virginia, 109, 111, 123
Speyer, Edmund, 123
Syrettes
  Second World War, 98, 99, 102
  Squibb Institute, 91

W
Whyte, William H.
  The Organizational Man, 100, 104
Wieland, Heinrich
  laboratory, 109
  treatment by the Nazis of, 111
Wiley, Harvey
  BOC, 79–81
  employed by *Good Housekeeping
      Magazine*, 86
Wood, Dr. Alexander
  hypodermic needle debate, 28, 63, 98

Printed in the United States
By Bookmasters